T0329087

CHARACTERIZATION AND TREATMENT OF TEXTILE WASTEWATER

CHARACTERIZATION AND TREATMENT OF TEXTILE WASTEWATER

HIMANSHU PATEL

R.T. VASHI

AMSTERDAM • BOSTON • HEIDELBERG • LONDON
NEW YORK • OXFORD • PARIS • SAN DIEGO
SAN FRANCISCO • SINGAPORE • SYDNEY • TOKYO

Butterworth-Heinemann is an imprint of Elsevier

Butterworth Heinemann is an imprint of Elsevier
225 Wyman Street, Waltham, MA 02451, USA
The Boulevard, Langford Lane, Kidlington, Oxford OX5 1GB, UK

Notices
Knowledge and best practice in this field are constantly changing. As new research and
experience broaden our understanding, changes in research methods, professional
practices, or medical treatment may become necessary.

Practitioners and researchers must always rely on their own experience and knowledge
in evaluating and using any information, methods, compounds, or experiments described
herein. In using such information or methods they should be mindful of their own safety
and the safety of others, including parties for whom they have a professional responsibility.

To the fullest extent of the law, neither the Publisher nor the authors, contributors,
or editors, assume any liability for any injury and/or damage to persons or property as
a matter of products liability, negligence or otherwise, or from any use or operation
of any methods, products, instructions, or ideas contained in the material herein.

Library of Congress Cataloging-in-Publication Data
A catalog record for this book is available from the Library of Congress

British Library Cataloguing in Publication Data
A catalogue record for this book is available from the British Library

For information on all Elsevier publications
visit our website at http://store.elsevier.com/

ISBN: 978-0-12-802326-6

CONTENTS

ACKNOWLEDGMENT

I take this opportunity to express my appreciation to all those who have contributed a lot in making this work possible and have also given me support in all the things that I needed.

I would like to thank Dr. Ashwin Patel, principal, and Dr. G. M. Malik, head of the Department of Chemistry, Navyug Science College, for providing me their laboratory facilities. I would also like to thank all teaching and nonteaching staff for their kind support during the course of my study.

I express my deep feeling of appreciation to Hashmukh C. Patel, manager of a well-known dyeing mill, who gave me information about the processes occurring in textile mills and also helped me in collecting wastewater. I am thankful to Dr. K. C. Patel, professor of the Department of Chemistry, Veer Narmad South Gujarat University, Surat, and to my friends Purvish Shroff, Ketan Patel, Sagar Desai, Nilesh Ahire, and Mukesh Tandel for giving me moral support and encouraging me during my nervous moments.

I owe my loving thanks to all my family members. Without their encouragement and understanding, it would have been impossible for me to finish this work. I am really staggered to realize just how much effort they have made and put in for me to reach this achievement.

I hereby accept with a deep sense of satisfaction and pleasure the obligation of the abovementioned and also those whom I may have failed to mention; without them, this endeavor would have been a distant goal.

Himanshu J. Patel
Department of Chemistry, Navyug Science College
Surat, Gujarat, India

CHAPTER 1

Introduction

Contents

Abstract

This chapter begins with an overview of the current environment system, including the definition of environment and pollution. Water pollution is created from various human activities like population explosion, haphazard rapid urbanization, and industrial and

Characterization and Treatment of Textile Wastewater
http://dx.doi.org/10.1016/B978-0-12-802326-6.00001-0

technological expansion producing effluent, in which the textile industry, one of the essential and largest sectors, consumes a large amount of water and generates a substantial quantity of water. Detailed general processes (dry and wet) of textile wastewater with their usage of chemicals with flowchart are exposed. Also, major pollutant/chemical types and effluent composition per processes were integrated, which indicates wastewater coming from textile processes is highly polluted. Traditional (primary, secondary, batch, and sludge conditioning) and advanced treatments (adsorption, coagulation, and neutralization) of wastewater for contamination removal are discussed. Remedial measurement and hazardous effect of wastewater, especially on biotic components like aquatic and terrestrial components, are mentioned, if it is disposed to the environment directly.

Keywords: Environmental chemistry, Pollution, Dyeing industrial process, Treatment and characterization of wastewater, Remedial measurements.

LIST OF FIGURE

LIST OF TABLES

1.1 CURRENT SCENARIO OF ENVIRONMENT

The environment is the sum of all social, economical, biological, physical, and chemical factors, which constitutes the surrounding of humans. The relationship of humankind with the environment is symbiotic. The environment is complex and dynamic in which all life forms are interdependent.[1–3] The environment performs three basic functions in relation to humankind. First, it provides living space and other amenities that make life qualitatively rich for humankind. Second, the environment is a source of agricultural, mineral, water, and other resources that are consumed directly or indirectly by humans. Third, the environment is a sink where all the waste produced by humans is assimilated. It is essential that the capacity of the environment to perform these functions is not impaired; thus, it explains our general concern for it. It is important, therefore, that due to stresses imposed on the environment, the rate of exploitation of resources does not exceed nature's capacity to reproduce them.

Environmental chemistry is the chemistry that generally focuses on the chemical phenomena in the environment. It deals with chemical composition, structure, properties, reactions, transport, effects, and fate of different chemical spices in the environment—air, water, soil, and their effects on living organisms particularly human beings. Humans live in the natural world of beautiful plants, animals, fresh air, clean water, and fertile soil that fulfill all their basic needs such as food, water, and shelter. Human populations and their activities, for example, industrialization, urbanization, and deforestation, have grown at alarming rates that in turn accelerate the extraction and modification of our environment to such an extent that threatens both our continued existence and that of many organisms. Humans have been continuously disturbing the delicate balance of nature and are changing the basic characteristics of the environment by removing some of its essential components; as a result, almost every aspect of modern living possesses a potential health risk. The environment encompasses everything that is around us, that is, air, water, and land. Air, water, and land have been contaminated with chemical additives called "pollutants."[4–7]

Pollution is the introduction of contaminants into an environment, of whatever predetermined or agreed upon proportions or frame of reference; these contaminants cause instability, disorder, harm, or discomfort to the physical systems or living organisms therein. Pollution of water may be defined as the addition of undesirable substances or unwanted foreign matter into the water bodies, thereby adversely altering natural quality of water. It is of vital concern to humankind, since it is directly linked with human welfare.[8]

Science and technology, as part of their contribution to economic and social development, must be applied to the identification, avoidance, and control of environmental risks and the solution of environmental problems and for the common good of humankind. Humans have the fundamental right to freedom, quality, and adequate conditions of life, in an environment of quality that permits a life of dignity and well-being, and they bear a solemn responsibility to protect and improve the environment for the present and future generations. In this respect, policies promoting or perpetuating apartheid, racial segregation, discrimination, colonial and other forms of oppression, and foreign domination stand condemned and must be eliminated.

Water is obviously an important topic in environmental science; it is a vitally important substance in all parts of the environment. Water covers about 70% of Earth's surface. It occurs in all spheres of the environment—in the oceans as a vast reservoir of saltwater, on land as surface water in lakes and rivers, underground as groundwater, in the atmosphere as water vapor, and in the polar ice caps as solid ice. Water is an essential part of all living systems and is the medium from which life evolved and in which life exists.[9] Continuous economic development and growth of the developed and developing countries around the world lead to a considerable increase of water demand. The worldwide demand for high-quality water resources will be difficult to meet in the foreseeable future because of dwindling supply. The imbalance in the demand and supply of water resources will become a major problem confronting every country in the incoming few decades.[10]

Water quality characteristics of aquatic environments arise from the multitude of physical, chemical, and biological interactions. Water bodies like rivers, lakes, and estuaries are continuously subject to a dynamic state of change with respect to their geologic age and geochemical characteristics. This is demonstrated by continuous circulation, transformation, and accumulation of energy and matter through the medium of living things and their activity. This dynamic balance in the aquatic ecosystem is upset by human activity, resulting in pollution that is manifested dramatically as fish kill, offensive taste and odor, etc. Water quality characterization must take into account (i) the distribution dynamics of chemicals in the aqueous phase (soluble, colloidal or absorbed, or particulate matter), (ii) accumulation and release of chemicals by aquatic biota, (iii) accumulation and release by bottom deposits, and (iv) input from land and atmosphere, that is, airborne contamination and land runoffs.[11]

According to the American Dye Manufacturers Institute (ADMI), the largest trade association for the industry, capital expenditure by domestic dyeing companies has increased in recent years reaching $2.9 billion in 1995.[12] Dyeing effluents contain several types of pollutants, such as dispersants, leveling agents, salts, carriers, acids, alkali, and various dyes; wastewater quality is variable and depends on the kind of process that generates the effluent. Most environmental concern relates to the effluents of the dyeing and finishing processes that contain a variety of contaminations of higher concentration of chemical oxygen demand (COD), biological oxygen demand (BOD), suspended solids, organic nitrogen, and some heavy metals. Color is usually noticeable at dye concentrations above 1 mg/L and has been reported in effluent from textile manufacturing at exceeding concentrations mainly because 10-15% of the dye is lost into wastewater during the dyeing processes.[13]

ATIRA's work in the field of water commenced in the recent years with water consumption surveys and has extended over the years to water treatment, water supply, and effluent treatment. Numbers of water-consuming machines have been studied by ATIRA, and the knowledge and experience gained from these studies indicate that one of the large consumers of water worldwide is the dyeing industry. The dyeing industry is one of the oldest and largest industries. Dyeing mills need large quantities of water for different purposes, and their demand is increasing every year with expansion and introduction of new finishing processes. On the other side, the industrial water supply position in the majority of textile centers is becoming more critical and water cost is rising steeply. Measures to conserve water, therefore, are of great importance. Apart from consuming large quantities of water, mills also discharge a variety of effluents. Before disposal, they need to be treated to certain tolerance limits, since pollution control is strict all over the world. The proper effluent treatment is receiving an increasing attention.[14]

1.2 GENERAL PROCESSES IN DYEING INDUSTRY

Dyeing effluent varies from day to day and even hour to hour due to the batchwise nature of the dyeing process and is therefore difficult to characterize. The composition is determined by the processes involved, fiber type, and chemicals used. The most pronounced variations include the color of the wastewater and the type of dye contained in it. The strong color of textile wastes is the hardest component to treat. The effluent typically contains a large number of compounds as demonstrated by one report that, on analysis of wastewater streams from four factories, positively identified 314 compounds, determined the partial structure of 94, and detected an additional 107 unknown compounds. The principal pollutants in textile effluent are aromatics, halogenated hydrocarbons, and metals.[15]

General processes involved in dyeing/textile are divided into two processes.[16–18]

1.2.1 Dry process

The dry process consists of (i) opening, blending, and mixing, (ii) carding, (iii) combing, (iv) spinning, (v) weaving, and (vi) knitting. Generally, in this process, negligible water is used.

1.2.2 Wet process

The wet process consists of (i) singeing, (ii) desizing, (iii) kiering, (iv) bleaching, (v) mercerizing, and (vi) dyeing. These processes are subjected to a series of operations, which require appreciable quantities of water at each stage. Figure 1.1 depicted the flowchart of general wet processes in textile industries.

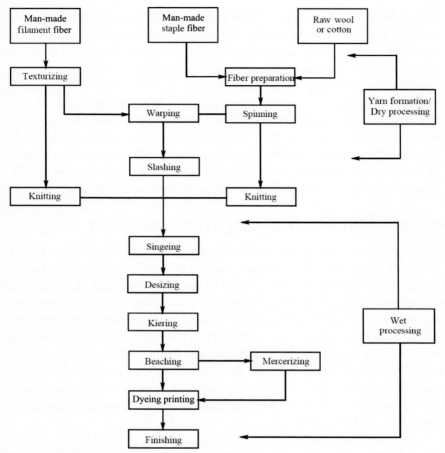

Figure 1.1 Flowchart of general textile wet processes.

Singeing

It is a continuous process in which the fabric travels near a series of jet burners at a faster rate through the machine in order to remove the protruding fibers from the surface. The flame burns off the fibers so as to render the surface of the fabric smooth, actually resulting in the burning of the infusible fibers to smoothen surface. Loading it with starch or other nonbiodegradable sizing agents strengthens the fiber. The wastewater from this process consists mostly of starch and softeners originating from the washing of various process vessels and sizing materials left over after the completion of operations. Carboxylmethyl cellulose (CMC) and polyvinyl alcohol (PVA) are also being used as substitutes for starch.

Desizing

After singeing, desizing is done to remove natural impurities and singeing compounds. Man-made fibers are generally sized with water-soluble sizes that are easily removed by a hot-water wash or in the scouring process. Certain types of enzymes are used to hydrolyze the starch. Removing starch before kiering is necessary because it can react and cause the color changes when exposed to sodium hydroxide in kier. Sometimes, acids may also be used for this purpose. Wastewater having high concentrations of organic material mainly consisting of breakdown products of sizing materials and agents used to bring about hydrolysis originates from desizing of fabric. Organic material is present in both dissolved and colloidal forms.

Kiering

The fibers from desizing may still contain grease, lubricants, antistatic agents, waxes, etc. They are removed by scouring with alkaline liquor containing caustic soda, soda ash, sodium silicate, sodium peroxide, etc., for several hours with the aid of stream. The entire operation is carried out on a batch basis, in which the spent liquor is blown out instantaneously.

Bleaching

The fabric after scouring is treated with bleach liquor. This liquor is mostly hydrogen peroxide. The other chemicals used are sodium chloride, formic acid, sulfuric acid, caustic soda, and hypochlorites to remove natural coloring material. Washing the fabric first with water and then with dilute acid and sodium bisulfate to remove last traces of chlorine and alkali follows bleaching. In the next step, the fabric is soaped, washed, and treated with optical whitening agent to improve whiteness of fabric. This operation contributes about 10-20% of the total pollution load. Hydrogen peroxide blending reduces the total pollution load in the wastewater causing no residual solids. The wastewaters from this unit contain soap and optical whitening agent.

Mercerizing

This process consists of treatment with a cold concentrated caustic alkali solution and washing with liberal amount of water in a countercurrent system. Mercerizing swells the fabric, imparting increasing dye affinity, tensile strength, and luster to the fabric.

Dyeing

The nature of wastewaters coming from this unit depends upon the use of different types of dyes and auxiliary chemicals. Classes of dyes include vat dyes, disperse dyes, acid dyes, basic dyes, direct dyes, reactive dyes, naphthol dyes, sulfur dyes, azoic dyes, developed dyes, fluorescent dyes, mordant dye, indigo dyes, and oxidation–based dyes. Dyeing section contributes 15-20% of the total wastewater flow. The wastewater is strongly colored, and the color changes too frequently because of the change in the types of dyes used.

Finishing

In this unit, the fabric is washed in an open soaping range to remove the unfixed dyes. Then, it is treated with starches to finish the fabric. The other materials may be dextrins, natural and synthetic waxes, and synthetic resins. Washing of the fabric to remove the unused color and cleaning of the color machine constitute major source of wastewater. The wastewater is strongly colored and contains fixing agent like gum, soap, and minerals.

1.3 SOURCES OF WASTEWATER

The water consumption varies widely in the industry depending on the mill, processes, equipment used, and types of materials produced. Each textile process utilizes a large amount of water, which will finally become wastewater. The most significant sources of pollution among various process stages are pretreatment, dyeing, printing, and finishing of textile materials. Desizing is the industry's largest source of pollution. During desizing, all the sizes used during weaving are removed from the fabric and discarded into the wastewater. In scouring, dirt, oil, and waxes from natural fibers are removed from the fabric and washed into wastewater stream. Normally, desizing and scouring are combined and these two processes may contribute to 50% of BOD in the wastewater in the wet processing. Pollution from the peroxide bleaching is not a major problem.

Dyeing wastewater generates the largest portion of the total wastewater. The source of wastewater is from the dye preparation, spent dye bath, and washing processes. Dyeing wastewater contains high salt content, alkalinity, and range of color. Finishing processes generate organic pollutants such as residue of resins, softeners, and other auxiliaries. A composite wastewater from an integrated textile plant consists of the following materials: starches, dextrin, gums, glucose, waxes, pectin, alcohol, fatty acids, acetic acid, soap, detergents, sodium hydroxide, carbonates, sulfides, sulfites, chlorides, dyes,

Table 1.1 Major Pollutant Types in Textile Wastewaters and Their Origin

Pollutants	Major chemical types	Main processes of origin
Organic load	Starches, enzymes, fats, greases, waxes, surfactants, acetic acid	Desizing, bleaching, dyeing
Color	Dyes, scoured wool impurities	Dyeing
Nutrients (N and P)	Ammonium salts, urea phosphate-based buffers, and sequestrants	Dyeing
pH and salt effects	NaOH, mineral/organic acids, sodium chloride, silicate, sulfate, carbonate	Scouring, desizing, bleaching, mercerizing, dyeing
Sulfur	Sulfate, sulfide, hydrosulfite salts, and sulfuric acid	Dyeing
Toxicants	Heavy metals, reducing agents (e.g., sulfide), oxidizing agents (e.g., chlorite, peroxide, dichromate, and persulfate), biocides, quaternary, ammonium salts	Desizing, bleaching, dyeing, finishing
Refractory organics	Surfactants, dyes, resins, synthetic sizes (e.g., PVA), chlorinated organic compounds, carrier organic solvents	Bleaching, desizing, dyeing, finishing

pigments, carboxymethyl cellulose, gelatin, peroxides, silicones, fluorocarbons, and resins.[19,20]

The major pollutant types identified in textile wastewater are summarized in Table 1.1 along with their main origin in the textile manufacturing process.[21]

1.4 CHARACTERISTICS OF WASTEWATER

The quantity of process wastewater from each unit operation in textile mill is relatively small as compared to the quantity derived from sizing and desizing operation, following bleaching, dyeing, and printing. The combined wastewater volume from Indian mills, which use raw water of about 61–646 L/kg of cloth, lies in the range of 86–247 L with an average of 172 L/kg of cloth processed, that is, 17–50 L with an average of 35 L/m of cloth. On an average, the wastewater discharge corresponds to 58–81% of water consumed with an average value of 73%, which agrees with the water used in wet processes.[21]

Textile processing involves many different steps, in which wastewater is generated. The amount and composition of these wastewaters depend on many different factors, such as the type of fabric, the type of process, and

Table 1.2 Effluent Characteristic from Textile Industry

Process	Effluent composition	Nature
Sizing	Starch, waxes, carboxymethyl cellulose (CMC), polyvinyl alcohol (PVA), wetting agents	High in BOD, COD
Desizing	Starch, CMC, PVA, fats, waxes, pectin	High in BOD, COD, SS, DS
Bleaching	Sodium hypochlorite, Cl_2, NaOH, H_2O_2, acids, surfactant, $NaSiO_3$, sodium phosphate, cotton fiber	High alkalinity, high SS (suspended solid)
Mercerizing	Sodium hydroxide, cotton wax	High pH, low BOD, high DS
Dyeing	Dyestuffs urea, reducing agents, oxidizing agents, acetic acid, detergents, wetting agents	Strong colored, high BOD, high DS, low SS, low heavy metals
Printing	Pastes, urea, starches, gums, oils, binders, acids, thickeners, cross-linkers, reducing agents, alkali	Highly colored, high BOD, oily appearance, High Suspended Solid, slightly alkaline, low BOD

used chemicals. A large amount of hazardous compounds is emitted from textile and dyeing industry. Although effluent characteristics differ greatly even within the same process, some general values can be given. The characterization of wastewater of individual process suggested that wastewaters are highly polluted including high concentration of organics, color, and metals. The nature of the processing exerts a strong influence on the potential impacts associated with textile manufacturing operations due to the different characteristics associated with these effluents, which is mentioned in Table 1.2.[22–24]

1.5 TREATMENT OF WASTEWATER

Environmental heritage is the birthright of all people, and, hence, all our efforts to mitigate environmental pollution are most necessary to hand over unpolluted environment to the next generation. There are still many industries, which discharge their effluent by dilution, disposing directly to streams or waste course. But due to the enactment of the Water (Prevention and Control of Pollution) Act 1974, all industries are installing their own treatment plants and most of them are in operation in South Gujarat region. The treatment of industrial wastewater is complicated by the presence of a wide variety of both inorganic and synthetic organic pollutants, many of which are not readily susceptible to biodegradation. Solvents, oils, plastics, metallic wastes, suspended solids, phenols, and various chemical derivatives of manufacturing process are apt to be difficult to identify and impossible to remove without more advanced technology than we now know. Some waste problems in nature are taken care of by the large amount of living space; unfortunately, where living space becomes limited and dilution is no longer a satisfactory solution, other means must be taken to reduce the ratio of waste to space.

The basic principles of industrial waste treatment are based upon (i) the separation of solids from the liquid, (ii) oxidation of organic and oxygen-demanding materials, (iii) neutralization, (iv) removal of poisonous sub-stances, and (v) disposal of residues.

The methods used are physical, chemical, and biological in nature or may constitute a combination of various methods. The methods used for the sep-aration of solids from the liquid are confined in most instances to the following:

(1) The removal of soluble and suspended solids may be enhanced by the application of coagulants and precipitant and by screening and sedimentation.

(2) The oxidation of organic materials is accomplished by biological methods. Aeration alone may have some value when volatile sub-stances are present, but it is rarely sufficient or effective.

(3) Large volumes of waste requiring neutralization are acidic in character. Neutralization with lime, sodium hydroxide, sodium carbonate, or combinations of neutralization agents commonly employed usually requires equalization for economy and control. Neutralization of alkaline waste with various acids is carried out under certain conditions.

(4) The removal of poisonous substances requires specific method of treat-ment, depending upon the type and amount of poisonous material present. Residue produced by the treatment of industrial wastes varies from thin slurries to precipitates. The disposal of the residues depends upon the character of the materials.[25–28]

A brief outline of the technology and methodology for the treatment of wastewater generated from known textile and dyeing units is placed at Pan-desara GIDC, Surat, Gujarat. The proposed technology consists of the treat-ment for organic contaminations, grease and oil, color, and heavy metals.

The methods of treatment used for altering the characteristic liquid wastes fall into the following classification:

(1) Primary treatment
(2) Secondary treatment
(3) Batch treatment
(4) Sludge conditioning

1.5.1 Primary treatments

Primary treatments include those processes that reduce the floating and suspended solid present in the waste by mechanical means or by the action of gravity. Fine screens and sedimentation tanks are commonly used in primary treatment processes. The raw effluent will first flow through a manually operated bar screen chamber for the removal of leaves, twigs, and larger size particles. Effluent also contains grease and oil, which produce foul odor when discharged in water, and it is not easily digested in sludge digestion tanks. These are removed in skimming tanks. From that, the effluent will flow to a flocculation tank where lime will be added for pH correction. Aluminum sulfate (alum) will also be added in flocculation tank to improve the settling nature of suspended solids.

The settling process is assisted by the addition of chemical coagulants, which are applied as components of various levels of wastewater treatment. Coagulants serve two principal functions to assist in the coagulation and flocculation processes, in order to maximize the removal of very small solid particles of various compositions. The primary treatment units will generally remove 98-99% settable solids, 60-80% suspended solids, and 30-50% of oxygen demand from the waste. Primary treatment in effect uses the force of gravity to separate the raw wastewater into a water component and a concentrated solid or a sludge component.[29,30]

1.5.2 Secondary treatments

Secondary treatment depends on biological processes to further reduce the suspended and dissolved solids, which are remaining in the liquid effluent after primary treatment. From the primary treatment, effluent water will be pumped by filter feed pump to pressure sand filter where suspended solids will be reduced prior to entry into the series of exchangers. Ion exchange is basically a process of exchanging certain cations and anions of the wastewater for sodium, hydrogen, or other ions in a resinous material.

The resins, both natural and artificial, are commonly referred to as zeolites. The ion-exchange process was originally developed to reduce hardness in domestic waste supplies but has recently been used to treat industrial wastewater, such as metal-plating waste for the softening of water. Organic matter and pH have a pronounced effect on the operation and efficiency of resin beds. Chemicals used for regenerating resin beds also require special treatment before disposal. Ion exchange also requires careful operation and supervision at all time. There are two types of exchangers used:

(1) Effluent passes through a strong acid cation exchanger whereby all the cations are replaced by the hydrogen ions.

(2) Effluent then passes through a weak base anion exchanger, where anions are replaced by hydrogen ions.

The water coming out from the anion exchanger will be in the form of demineralized water and can be reused in the process.

1.5.3 Batch treatment

Regeneration and rinse water containing cations will be collected in a batch treatment tank. Alkaline chemical (lime or caustic soda) is added when the tank is full and a mixer will mix the contents well. After mixing, the contents will be allowed to settle, and then, the water above the sludge will be decanted to the drain. Finally, the sludge pump will draw off the sludge remaining.

1.5.4 Sludge conditioning

The production of metal finishing sludge is the result of the widespread use of chemical waste treatment in the industry. This method of waste treatment involves the reduction of chromates, oxidative destruction of cyanides, and the precipitation of heavy metals as hydroxides. The metal-bearing sludge produced is difficult to handle and the final disposal of these solid wastes is a troublesome and costly aspect of chemical waste treatment. In addition, the toxic heavy metals contained in this sludge have the potential for causing harmful environmental effects. Waste recovery techniques, which have achieved commercial status to date in the metal finishing industry, are ion exchange and electrolytic bath purification. Evaporation, ion exchange, reverse osmosis, and electrodialysis for rinse waters are used to alleviate this burdensome sludge problem.[31,32]

1.5.5 Physico-chemical treatments (advanced treatment)[33-38]
Adsorption

Adsorption is a surface phenomenon that is defined as the increase in the concentration of a particular component at the surface of interface of the two phases. Adsorption can be (i) physical adsorption (physisorption), which involves relatively weak intermolecular forces, and (ii) chemisorption, which involves essentially the formation of a chemical bond between the sorbent molecule and the surface of the adsorbent.

Physisorption is a general phenomenon with a relatively low degree of specificity, where chemisorption is dependent on the reactivity of the adsorbent and adsorptive. Chemisorbed molecules are linked to reactive parts of the surface, and the adsorption is necessarily confirmed to a monolayer. At high relative pressures, physisorption generally occurs as a multilayer. A physisorbed molecule keeps its identity and on desorption returns to the fluid phase in its original form. If a chemisorbed molecule undergoes reaction or dissociation, it loses its identity and cannot be recovered by desorption. The energy of chemisorption is the same order of magnitude as energy change in a comparable chemical reaction. Physisorption is always exothermic, but the energy involved is generally not much larger than the energy of condensation of the adsorption. However, it is appreciably enhanced when physisorption takes place in very narrow pores. An activation energy is often involved in chemisorption, and at low temperature, the system may not have sufficient thermal energy to attain thermodynamic equilibrium. Physisorption systems generally attain thermodynamic equilibrium. Physisorption systems generally attain equilibrium fairly rapidly, but equilibration may be slow if the transport process is rate-determining. Large surface area is required on the adsorbent where small molecules of low molecular weight adhere to each other in a square meter surface. In a solid–liquid system, adsorption takes place with the result of the removal of solute from solution. The concentration of solute on the surface of solid continues to increase until solute in solution remains in equilibrium with that at the surfaces.

Coagulation

The term "coagulation" comes from the Latin coagulare meaning to drive together. This process describes the effect produced by the addition of a chemical to a colloidal dispersion resulting in particle destabilization by a reaction of the force tending to keep the particle apart. Coagulation is achieved by adding the appropriate chemical, which causes particles to stick together when contact is made. The entire process occurs in a very short time, probably less than a second, and initially results in particles submicroscopic in size. Coagulation in water treatment is invariably achieved by using salts, which hydrolyze in water. Aluminum salt or iron salt is added to turbid water.

When alum is added to water during coagulation, the pH of the water is depressed due to the presence of excess hydrogen ions. Hydrolyzing metal ions has developed significant insights for the mechanisms of coagulation over the last decade. The coagulation reactions are an interaction between the hydrolysis product of Al(III) and a colloidal suspension. The kinetics of the reactions of coagulation in water treatment with aluminum salts occurs predominantly by two mechanisms: adsorption of the soluble hydrolysis species on the colloidal and destabilization. These processes are extremely fast and occur within microseconds. The settling velocities of finely divided and colloidal particles under gravity alone are so small that ordinary sedimentation is not practical. It is necessary, therefore, to use procedures that agglomerate the small particles into larger aggregates, which then have the settling velocities required for practical purposes. The formation of larger particles from smaller ones is also required for their removal by filtration.

Optimum coagulation treatment of raw water represents the attainment of a very complex equilibrium in which many variables are involved. This reaction results in the formation of both monomeric and polymeric metal hydroxide species that are important in the coagulation of turbidity-causing materials in water. Like other chemical reactions, it is possible that both the kinetics and equilibrium of the metal hydroxide precipitation are affected by changes in temperature.[39–42]

Neutralization

The neutralization of an acid with a base appears to be simple, yet in the actual practice of acid-waste neutralization, many difficulties arise during neutralization. The neutralization agents are sodium hydroxide, sodium carbonate, ammonia, quick lime, hydrated lime, and limestone in various grades. Lime and limestone are available as high calcium or dolomitic materials. Sodium hydroxide, sodium carbonate, and ammonia are satisfactory neutralization agents, but on account of their cost, these neutralization agents are seldom used. The treatment of metals, oils, and grease-bearing wastes by neutralization and precipitation usually involves recombining the waste with metal ions. The flux produced is large and heavy, and hence, the waste is then allowed to settle.

There are many acceptable methods for neutralization, such as mixing wastes, passing acid wastes through limestone beds, mixing acid wastes with lime slurries or dolomite lime slurries, and adding the proper proportions of concentrated solutions of caustic soda (NaOH) or soda ash (Na_2CO_3) to acid wastes. The addition of caustic soda or sodium carbonate to acid wastes in the proper proportions results in faster but more costly neutralization. Smaller volumes of these agents are required, since it is more powerful than lime or limestone. The choice of the best neutralizing agent in any instance will be governed by the economic position of a given plant waste treatment practice.

Acidic or alkaline wastes should not be excessively discharged without treatment into a receiving stream. At low or very high pH, water can adversely affect the aquatic life. The adverse condition is even more critical when sudden slugs of acid or alkali are imposed upon the stream.[43–47]

1.6 REMEDIAL MEASURES

Pollution control is an essential task. There are four types of control: legal, social, economical, and technological measures, which help to prevent the pollution by various methods of operations. Waste products enter the environment in various forms and threaten the quality of the air, land, and water. The presence of waste products in water is especially serious, as many of these products can enter the food chain, where the biochemical processes can rapidly increase their concentration to toxic level. Hence, it is extremely important to study the methods of treating waste products and eliminating them from aqueous system. The US Environmental Protection Agency has listed copper as a priority pollutant.

Pollution control has almost become an integral part of the process of industrialization. Appropriate laws have been passed that restrict and regulate the growth of pollution intensive industries, especially in metropolitan cities. It has been made obligatory for industrial units to adopt measures to control pollution.

The pollution powers of plating wastes are reduced within the plant by several means. Many recommendations for modifications in design and operation to reduce wastes have been suggested. The Ohio River Valley Water Sanitation Commission has published a guide for these practices. Additional modifications include the following:

(1) Installing a gravity-fed, nonoverflowing emergency holding tank for toxic metals and their salts

(2) Eliminating breakable containers for concentrated materials

(3) Designing special drip pans, spraying rinses, and shaking mechanism

(4) Reducing spillages, drag-out, and leakage to the floor or other losses by curbing the area and discharging these losses to a holding tanks

(5) Using high-pressure fog rinses rather than higher volume water washes

(6) Reclaiming valuable metals from concentrated plating bath wastes

(7) Evaporating reclaimed wastes to desired volume and returning to plating bath at rate equals to loss from bath

(8) Recirculating wet washer wastes from fine scrubbers[48–51]

1.7 EFFECT OF WASTES

The effect of dyeing waste may be conveniently divided into four fields:

(1) Toxicity to fish and fish food

(2) Effect on human beings

(3) Effect on sewers

(4) Effect on sewage treatment processes

1.7.1 Toxicity to fish and fish food

The toxic effects on fish and fish food of some of the chemicals that may be found in dyeing wastes are size, age, hardness, and species of aquatic organisms. The pH, temperature, hardness, total alkalinity, oxygen content, and the other dissolved substances in the receiving water are also of importance. In addition, the type and amount of stream biota, the degree and nature of other pollution sources, the extent of stratification, the amount of aeration, and the presence of synergistic or antagonistic compounds in the water must also be taken into account. The contamination of dye in public streams may present a risk to humans due to the aquatic living organisms such as fish that can accumulate dye into their tissues, affecting the food web.

1.7.2 Effect on human beings

Hazardous wastewater in the textile industries is that that is likely to be potentially hazardous for human beings. Such waste includes the following:

(1) Toxic or poisonous substances that cause adverse effects by inhalation or ingestion

(2) Corrosion agents that affect the body tissues

(3)   Irritation and/or inflammation of the tissue caused by certain chemicals
(4)   Chemicals that may cause allergy
(5)   Mineral oils and organic solvents that are inflammable
(6)   Carcinogenic, mutagenic, and teratogenic substances

Further, color can be considered as the earliest pollutant to be detected in polluted water. The extensive use of dyes, in both dye-manufacturing and dye-consuming industries, creates significant problems due to the discharge of colored wastewater. The presence of very small amounts of dyes in water (<1 ppm for some dyes) is highly visible and affects the quality of water bodies. Besides the effect to the environment, dyes can also cause deterioration in humans' health. Some dyes are found to be toxic, mutagenic, and carcinogenic. Dyes released by the industries can get into the water bodies and eventually contaminate the water supply system. Consumption of dye-polluted water can cause allergy reactions, dermatitis, skin irritation, cancer, and mutation in both babies and grown-ups.

1.7.3 Effect on sewers

Acidic wastes produced in dyeing mill processes are corrosive and will thus attack metal and concrete structures. Acids convert the soap in sewage to fatty acids forming floating scum, which clings to objects and which leads to sludge-dewatering difficulties because of the production of a sticky sludge.

1.7.4 Effect on sewage treatment processes

Dyeing wastes have deleterious effect on sewage treatment processes because of the toxic reaction of the waste chemicals on the biological organisms.[52–55]

REFERENCES

1. Inderjeet S, Sethi MS, Iqbal SA. *Environmental pollution: cause, effects and control.* New Delhi: Common Wealth Publishers; 1991.
2. Manahan SE. *Environmental chemistry.* 3rd ed. Boston, MA: Willard Grant Press; 1979.
3. Fuller ED. *Chemistry and man's environment.* Boston, MA: Houghton Mifflin Co.; 1974
4. De AK. *Environmental chemistry.* 2nd ed. New Delhi: Wiley Eastern; 1989.
5. Thakur AK. *Water pollution.* Ph.D. Thesis, Muzaffarpur: University of Bihar; 1978.
6. Liptak BG, editor. *Water pollution. Environmental engineer's handbook,* vol. 1. Radnor, PA: Chilton Book; 1974.
7. Pani B. *Textbook of environmental chemistry.* New Delhi: I.K. International Publishing House Private Limited; 2007.
8. Vershney CK, editor. *Water pollution and management.* New Delhi: Wiley Eastern; 1985.

9. Nagrajan N. *Industrial safety and pollution control handbook.* Hyderabad: Joint Publication of National Safety Council and Associate (Data) Publishers; 1993.
10. Lin SH, Chen ML. Combined ozonation and ion exchange treatments of textile wastewater effluents. *J Environ Sci Health Part A* 1997;**A2**(7):1999–2010.
11. De AK. *Environmental chemistry.* 6th ed. New Delhi: New Age International Publishers; 2006.
12. ADMI. *Textile dyeing wastewater: characterization and treatment.* American Dye Manufacturers Institute, Rep. No. EPA 600/2-78-098, Washington, DC: U.S. Environmental Protection Agency; 1978.
13. Marcucci M, Ciabatti I, Mattecci A, Vernaglione G. Membrane technologies applied to textile wastewater treatment. *Ann NY Acad Sci* 2003;**984**:53–64.
14. ATIRA. *Water supply and consumption in textile mills.* An ATIRA survey report, 1968.
15. Isık M, Sponza DT. Anaerobic/aerobic treatment of a simulated textile wastewater. *Sep Purif Technol* 2008;**60**(1):64–72.
16. Rajagopalan S. *Water pollution Problem in Textile Industry and control.* In: Pollution Management in Industries. Ed. R. K. Trivedy. Environmental Pollution, Karad, India; 1990.
17. Pandey GN, Carney GC. *Environmental engineering.* New Delhi: Tata McGraw-Hill Publishing; 2008.
18. Bisschops I, Spanjers H. Literature review on textile wastewater characterization. *Environ Technol* 2003;**24**(11):1399–411.
19. Board N. *The complete technology book on textile spinning, weaving, finishing and printing.* New Delhi: Asia Pacific Business Press; 2003.
20. Noyes R. *Pollution prevention technology handbook.* NJ, USA: Noyes Publications; 1993.
21. Delee W, O'Neill C, Hawkes FR, Pinheiro HM. Anaerobic treatment of textile effluents: a review. *J Chem Technol Biotechnol* 1998;**73**:323–35.
22. Bal AS. Wastewater management for textile industry—an overview. *Indian J Environ Health* 1999;**41**(4):264–90.
23. Yusuff RO, Sonibare JA. Characterization of textile industries effluents in Kaduna, Nigeria and pollution implications. *Global Nest Int J* 2004;**6**(3):212–21.
24. Cervantes FJ, Pavlostathis SG, Van Haandel AC, editors. *Advanced biological treatment processes for industrial wastewaters.* London, UK: IWA Publishing; 2006, p. 267–97. ISBN: 1843391147.
25. Trivedi PR, Raj G. *Environmental industrial pollution control.* 1st ed. New Delhi: Akashdeep Publishing; 1992, 3 & 4.
26. Besselievre EB. *Treatment of industrial waste.* New York: McGraw Hill Publication; 1969.
27. Bond RG, Scraules BP. *Handbook of environmental pollution control. Wastewater treatment and disposal.* Cleveland, OH: CRC Press; 1979.
28. Parran T. Public health service and industrial pollution. *Ind Eng Chem* 1947;**39**(5):560–1.
29. Modak NV. *Sewage and wastewater treatment.* Bombay: Somaily Publication; 1971.
30. Nemerow NL. *Theories and practices of industrial waste treatment.* New York: Additional Wiley; 1963.
31. Cherry RH, Smithson GR. *Ciancia J Reclaimation of Metal value from residues produced during the treatment of metal finishing wastewaters,* Proceedings of the National Conference on Management and Disposal of Residues from the Treatment of Industrial wastewaters. Washington DC: US EPA; 1975.
32. Weiner RF. Acute problems in effluent treatment. *Plating* 1967;**54**:1354–6.
33. Faust SD, Aly M. *Adsorption processes for water treatment.* Stoneham, MA: Butterworths; 1987.
34. Ruthven DM. *Principles of adsorption and adsorption processes.* New Delhi: Wiley; 1984.
35. Langmuir IJ. Chemical reactions at low pressures. *J Amer Chem Soc* 1915;**37**(5):1139–67.
36. Freundlich H. *Colloidal and capillary chemistry.* London: Methuen and Company; 1926.

37. Weber WJ, Jr. *Adsorption in physico-chemical process for water quality control.* New York: Wiley; 1972.
38. Rouquerol F, Rouquerol J, Sing K. *Adsorption by powders & porous solids: principal, methodology and applications.* London, UK: Academic Press; 1999.
39. Markrle S. Mechanism of coagulation in water treatment. *Proceedings of American Society of Sanitary Engineering (SAU). J Sanit Eng Div* 1962;**88**(SA 3):1962;1.
40. O'Melia CRA. Review of the coagulation process. *Public Works* 1969;**100**(5):87–98.
41. Amirtharajah A, Mills KM. Rapid-mix design for mechanisms of alum coagulation. *J Amer Water Works Assoc* 1982;**74**(4):211–6.
42. Hahn HH. *Effect of chemical parameters up on rate of coagulation.* Doctored dissertation, Cambridge, MA: Harvard University; 1968.
43. Gehm HW. Up-flow neutralization of acid wastes. *Chem Metal Eng* 1944;**124**.
44. Rudolfs W, Rudolfs E, Jr. Effect of dilution on sludge formed by neutralizing acid wastes with lime. *Public Works* 1944;**75**:24–8.
45. Dickerson BW, Brooks RM. Neutralization of acid wastes. *Ind Eng Chem* 1950;**42**:599–605.
46. Jacobs HL. Neutralization of acid wastes. *Sewage Ind Wastes* 1951;**23**:900–5.
47. Dillon KE. Waste disposal made profitable. *Chem Eng* 1967;**74**:146.
48. Ohio River Valley Water Sanitation Commission. *Plating room controls for pollution abatement*; 1951.
49. Duke K. Hazardous waste minimization: is it taking root in U.S. industry? *Waste Manage* 1994;**14**:49–59.
50. O'Shaughhnessy J, Clark W, Lizotte RP, Jr., Mekutel D. *Pollution prevention and water conservation in metals finishing operations. 50th Purdue Industrial Waste Conference, Purdue Univ., Lafayette, Indiana*; 1995, p. 735.
51. Keith LH, Telliard WA. Priority pollutants I: a perspective view. *Environ Sci Technol* 1979;**4**:13.
52. Rudolfs W, Barnes GE, Edwards GP, Heukelekian H, Hurwitz E, Renn CE, et al. Review of literature on toxic materials affecting sewage treatment processes, streams and B.O.D. determinations. *Sewage Ind Wastes* 1980;**22**(9):1157–91.
53. Tsat C. Effect of chlorinated sewage effluents on fishes in upper Patuxent river, Maryland. *Chesapeake Sci* 1968;**9**(2):86–93.
54. Banat IM, Nigam P, Singh D, Marchant R. Microbial decolourization of textile dyes containing effluents: a review. *Bioresour Technol* 1996;**58**:217–27.
55. Mckay G. Two-resistance mass transfer models for the adsorption of dyestuffs from solutions using activated carbon. *J Chem Technol Biotechnol* 1984;**34A**:294–310.

CHAPTER 2

Characterization of Textile Wastewater

Contents

Characterization and Treatment of Textile Wastewater
http://dx.doi.org/10.1016/B978-0-12-802326-6.00002-2

Abstract

In this chapter, the collection, preservation, analysis method, and report of physico-chemical and microbiological parameters of textile wastewater are mentioned. Principle, apparatus, interface, limitations, pretreatment, and procedure in short and detailed forms of 22 parameters are summarized. Water quality parameters are divided mainly into two parts: (i) titrimetric method [pollution-indicating parameters like chemical oxygen demand (COD) and biochemical oxygen demand (BOD), color, total hardness, magnesium hardness, and calcium hardness] and (ii) instrumentation parameters (pH, color, metals, etc.). The analysis report of 22 parameters for combined textile wastewater before treatment, collected from July 2009 to July 2010 bimonthly at GIDC Pandesara, Surat, Gujarat, is included. These parameters are compared with permissible limits of parameters for textile wastewater given by various firms including Central Pollution Control Board and US Environmental Protection Agency, which shows that COD and BOD values are 5 and 10 times, respectively, higher than limits. Characteristics of textile wastewater that were investigated and analyzed by different scientists are well fitted with one another. It shows that textile wastewater is highly polluted and treatment is necessary prior to discharge of it in environment.

Keywords: Combined wastewater, Collection, Preservation, Titrimetric and instrumental method, Characterization.

LIST OF TABLES

Table 2.1 Characteristics of Six Samples of Combined Wastewater of the Dyeing Mill Under Investigation
Table 2.2 Permissible Limits of Parameters for Textile Wastewater Given by Various Firms
Table 2.3 Characteristics of Effluents in M/s. Sivasakthi Textile Processors, Tirupur
Table 2.4 Physicochemical Characteristics of Composite Wastewater of Small-Scale Textile Industry
Table 2.5 Physicochemical Characteristics of Effluents from the Textile Mills

The dyeing industry is one of those industries that consume a considerable amount of water in the manufacturing process. Primary water is employed in the dyeing and finishing operations; dyestuffs used in the operation can vary from day to day and sometimes even several times a day mainly because of the batchwise nature of the dyeing process. The textile industry is generally concentrated in South Gujarat area covering metro cities. The dyeing house needs large quantities of water for different purposes, and their demand is increasing every year with expansion of new finishing processes. Raw materials for textile industry are mainly cotton, wool, and synthetic fibers, which are mainly from South Gujarat area. The synthetic fiber process unit requires large volumes of freshwater of fairly high purity. Water is used in these units for various operations such as steam generation, cooling, and demineralization. The overall consumption of water in a mill depends on the quantity and quality of cloth processed and the number of sequences adopted for rinsing and washing.[1,2]

Combined wastewater volume from Indian mills, which use raw volume of about 61–646 L/kg of cloth, lies in the range of 86–247 L with an average of 172 L/kg of cloth processed, thus 17–50 L with an average of 35 L/m of cloth. On an average, the wastewater discharge corresponds to 58–81% of water consumed with an average value of 73%, which agrees with the water used in wet processes. In these wet processes, dyeing process is more reliable for characterizations because the nature of wastewater coming from this unit depends on different types of dyes and auxiliary chemicals, which is dependent upon fibers used. In the dyeing mill, a variety of chemicals such as enzymes, acids, alkalis, hypochlorites, peroxides, and dyes, namely, direct, basic, vat, sulfur, naphthol, and metal complex, are used. Thus, the generated wastewater carries appreciable quantities of these chemicals along with other organic materials such as fats, waxes, pectins, solid fragments, and starches, derived from the cloth during its processing, which constitute a major source of pollution. Synthetic fiber processing unit involves mainly the wet processing operations like slashing, desizing, scouring, bleaching, mercerizing, and dyeing. There is no remarkable difference in water consumption between continuous and batch processes, if wet processing is observed. There are various chemicals in the processes: (i) starch, (ii) enzymes, (iii) acids, (iv) caustic soda, (v) soda ash, (vi) detergents, (vii) peroxides, (viii) hypochlorites, etc.[3,4]

2.1 METHODS

Methods of analysis employed are mostly the procedures described in the standard methods for water and wastewater.[5,6]

2.2 COLLECTION AND PRESERVATION OF SAMPLES

It is an old axiom that the result of any testing method can be no better than the sample on which it is performed. Collection and preservation of samples are almost important. It is essential to protect samples from changes in composition and deterioration with aging due to various interactions. Preservation of samples is essential for regarding biological action, hydrolysis of chemical compounds and complexes, and reduction of volatility of constituents. The objective of sampling is to collect a portion of material small enough in volume to be transported conveniently and yet large enough for analytic purposes while still accurately representing the material being sampled. In preserving samples, plastic-sampling bottles should be avoided when a reaction is possible between constituents of the waste, such as organic solvents and plastic. Likewise, metal containers and caps should not be used to hold wastes on which metals are to be determined. The bottles (polythene and glass) for sample collection should be thoroughly rinsed by repeated washing with deionized distilled water. They should be rinsed three times with the sample water before collection.[7] Dyeing mill wastewater was collected from June 2009 to June 2010. Composite samples are taken from the equalization tank, where effluent streams from various wet processing sections of the mill are collected and combined. Sampling is done bimonthly. To obtain representative samples, sampling is done after the shutdown of mill and mixing of effluent in the equalization tank. Sampling is carried out in a 10 L plastic container. The analysis of the effluent is performed within 24 h. Adsorption and coagulation studies are carried out within 48 h after sampling.

The analysis of quality (physicochemical and microbiological) of wastewater is the prime consideration to assess its effect on the entire ecosystem. Industrial wastewater is analyzed to decide upon what physical, chemical, or biological treatment should be given to make them suitable. Consider all treatments to find suitability. Various important parameters and methods in general practice are given below.

(1) Color
(2) pH
(3) Electrical conductivity (EC)
(4) Chemical oxygen demand (COD)
(5) Biochemical oxygen demand (BOD)
(6) Hardness
(7) Oil and grease
(8) Chloride
(9) Phenol
(10) Total dissolved solids (TDSs)
(11) Total alkalinity
(12) Fluoride
(13) Sulfate
(14) Phosphate
(15) Silica
(16) Sodium
(17) Heavy metals (Cu, Pb, Mn, and Cd)

Quantitative analytic procedures applied for analysis can be classified into three groups: (i) gravimetric, (ii) volumetric, and (iii) microanalytical analysis using sophisticated instruments, if necessary.

(i) *Gravimetric analysis*: methods are used on weight difference observed in sample by evaporation, filtration, or precipitation. This procedure is time-consuming, and a more precise chemical balance weighing up to 0.0001 g is necessary. This analysis included TDS and phenol.

(ii) *Volumetric analysis*: this method depends on the measurement of volumes of reagent of known strength. Normality of standard solutions are chosen in such a fashion that the titer value directly gives the value that represents the quantity of substance in the water pollution studies such that alkalinity, hardness, chloride, COD, and BOD are analyzed by using this analytic method.

(iii) *Microanalysis*: for this analysis, modern instrument-based techniques are used. The techniques like spectrophotometry, based on color intensity measurements, are most widely used to determine the concentration of the unknown substances for quantitative- and qualitative-specific determinations of recently sophisticated instruments like pH meter, atomic absorption spectrophotometer (AAS), flame photometer, gas chromatography, and electrodes, which are useful to analyze color, pH, EC, fluoride, sulfate, phosphate, sodium, and heavy metals (Cu, Pb, Mn, and Cd).

2.2.1 Color

Color in water may result from the presence of natural metallic ions (iron and manganese), humus and peat materials, plankton, and weeds; for example, iron oxides cause reddish water and manganese oxides cause brown or brackish water. Industrial wastes from textile and dyeing operation, pulp and paper production, food processing, mining and coal processing operation, refinery, and slaughterhouse operation may add substantial coloration to water in receiving stream. True color is the result of dissolved organics, minerals, or chemicals like dye in water, as noted above. Color is removed to make water suitable for general and industrial applications. Colored industrial wastewater may require removal of color before discharge into creek because small amount of color can be visible as contamination of water and causes esthetic and health problems. Colorless water is considered pure though it may be unsafe for human health. Generally, colored water imparts adverse effects on human health and aquatic environment. Highly colored water has significant effects on aquatic plants and algal growth. Light is very critical for the growth of aquatic plants, and colored water can limit the penetration of light. Thus, a highly colored body of water could not sustain aquatic life, which could lead to the long-term impairment of the ecosystem.

Principle

The color of a filtered sample is expressed in terms of the sensation realized when viewing the sample. These values are best determined from the light transmission characteristics of the filtered sample by means of a spectrophotometer.

Procedure

Measure color at 340 nm wavelength on spectrophotometer (Boss and Lumps) by preinstalled program.

Apparatuses

a. Spectrophotometer (Boss and Lumps)
b. For filtration: 100 mL Nessler tube, glass funnel, and 100 mL glass beaker
c. For dilution: 5, 10, and 25 mL volumetric pipettes and 100 mL volumetric flask
d. 10 mm quartz tube (cell)

Actual process

About 50 mL of thoroughly mixed homogeneous sample is taken out in 100 mL glass beaker. It is filtered through Whatman filter paper No. 1, and filtrate is collected in a 100 mL Nessler tube. Spectrophotometer is switched "on." While running, it first autochecks itself as such program is inbuilt. The 10 mm quartz cell is filled with the sample after rinsing it 2–3 times with sample. Cell is placed on a suitable place in the instrument. Test No. 25 measures color in Hazen units in the range of 1–500 at 340 nm. For samples having color higher than the prescribed range, dilution practice is used (for common effluent treatment plant and river water samples). Appropriate amount of sample is taken by means of pipette in a volumetric flask and the rest of volume is made up by distilled water. The dilution factor is installed in the instrument memory. Hence, it gives the color value directly. Distilled water blank is measured and deducted from the sample reading.[6,8]

2.2.2 pH

The basic principle of electronic pH measurement is the determination of activity of hydrogen ions by potentiometer measurement using a standard sensing electrode (glass electrode) and a reference electrode (calomel electrode). pH is a measurement of acid–base equilibrium achieved by various dissolved compounds. In most natural waters, pH is controlled by CO_2, CO_3, and HCO_3 equilibrium system. The pH of the environment has a profound effect on the rate of microbial growth. pH affects the function of metabolic enzymes. Acidic conditions (low pH) or basic conditions (high pH) alter the structure of the enzyme and stop growth. Most microorganisms do well within a pH range of 6.5–8.5. However, some enzyme systems can tolerate extreme pH and will thrive in acidic or basic environments. Extreme pH levels, the presence of particulate matter, the accumulation of toxic chemicals, and increasing alkalinity levels are common problems in wastewater.

Principle

A glass electrode (sensing electrode) consists of a thin glass bulb (special quality) containing a fixed concentration of HCl solution, into which a Ag–AgCl wire is inserted, serving as the electrode with a fixed voltage. When glass electrode is immersed in a solution, a potential difference develops between the solution in the glass bulb and sample solution. The potential difference E is formulated by the Nernst equation.

$$E = (RT/nF)\log(K/M^{n+}) \tag{2.1}$$

where E is the half-cell potential, R the gas constant, T the absolute temperature, n the valence, F is Faraday constant, K the constant, and M is the activity of ions to be measured.

The E (half-cell potential) cannot be measured alone. If the glass electrode is placed against a reference electrode, the potential difference between two electrodes is measurable $(E - E_{cal})$. Before any pH measurement, two electrodes have to be placed first in a solution of known pH (e.g., H^+ concentration $= 1$ g/L). This is called standardization of electrode and pH meter.

By definition, pH is the negative logarithm of hydrogen ion concentration, more precisely hydrogen ion activity:[9]

$$pH = -\log_{10}[H_3O^+] \quad \text{or} \quad pH = -\log_{10}\{1/[OH^-]\} \tag{2.2}$$

Apparatuses

a. pH meter (Equiptronics; Model: EQ-614 A)
b. 250 mL glass beaker

Reagents

i. Buffer solutions of known pH are used to standardize the pH meter.
ii. *Phthalate buffer (pH 4.0 at 25 °C)*: 10.12 g of potassium hydrogen phthalate ($KHC_8H_4O_4$) in 1 L distilled water.

iii. *Phosphate buffer (pH 7.0 at 25 °C)*: 3.4 g of monopotassium phosphate (KH_2PO_4) + 4.45 g of sodium hydrogen phosphate ($Na_2HPO_4 \cdot 2H_2O$) in 1 L fresh distilled water.

iv. *Borax buffer (pH 9.18 at 25 °C)*: 3.18 g sodium tetraborate decahydrate ($Na_2B_4O_7 \cdot 10H_2O$) in distilled water.

These pH standards are commercially available in the market and used as is (Merck, India).

Actual process

The pH meter is calibrated using standard buffer solutions of pH 4.0, 7.0, and 9.18 at room temperature. The pH of water samples is determined at room temperature. The electrode is washed thoroughly by distilled water and cleaned by filter paper before each measurement of sample and for buffer solutions. About 100 mL of thoroughly mixed homogeneous sample is taken out in a 250 mL glass beaker. The electrode is dipped in the sample. Instrument gives direct measurement of pH.[7]

2.2.3 Electrical conductivity

EC is a numerical expression of the ability of an aqueous solution to carry an electric current. This ability depends on the presence of ions, their total concentration, mobility, valence, and relative concentration and on the temperature of measurement. Conductivity in water is affected by the presence of inorganic dissolved solids such as chloride, nitrate, sulfate, and phosphate anions (ions that carry a negative charge) or sodium, magnesium, calcium, iron, and aluminum cations (ions that carry a positive charge). Further, EC of water is actually a measure of salinity. Excessively high salinity can affect plants, increasing specific toxicity of a particular ion (such as sodium), and also, higher osmotic pressure around the roots prevents an efficient water absorption by the plant. Organic compounds like oil, phenol, alcohol, and sugar do not conduct electric current very well and therefore have a low conductivity when in water. Conductivity is also affected by temperature: the warmer the water, the higher the conductivity. For this reason, conductivity is reported at 25 °C. Conversely, molecules of organic compounds that do not dissociate in aqueous solution conduct a current very poorly.

Principle

The physical measurement made in a laboratory determination of conductivity is usually the resistance measured in ohms or megohms. The resistance of a conductor is inversely propositional to its cross–sectional area and directly proportional to its length.

Conductance C is defined as the reciprocal of resistance R:

$$C = 1/R \tag{2.3}$$

where R is in ohm.

Further, the conductance (C) of a solution is directly proportional to the surface area (A, cm^2) and inversely proportional to the distance between the electrodes (L, cm):

$$C \propto A/L \tag{2.4}$$

$$C = kA/L \tag{2.5}$$

where k, the constant of proportionality, is called "conductivity." So

$$k = (C \times L)/A \tag{2.6}$$

Instrument
Digital conductivity meter (Equiptronics; Model: EQ-660A).

Apparatus
a. 100 mL glass beaker

Reagent
i. *Standard KCl solution (0.01 M)*: dissolve 745.6 mg anhydrous KCl in distilled water and dilute up to 1 L at 25 °C. It has a conductivity of 1413 $\mu\mho$/cm, for calibration of instrument.

Actual process
The conductivity cell is washed thoroughly by distilled water and cleaned by filter paper before each measurement for sample and for KCl standard solution. All measurements of conductance are made at 25 ± 0.1 °C temperature. For calibration, the conductivity cell is immersed into the standard KCl solution. Turn the knob to an appropriate range from $\mu\mho$ to m$\mho$. For sample measurement, 100 mL of thoroughly mixed homogeneous sample is taken out in 100 mL glass beaker. The conductivity cell is dipped in the beaker and EC is noted.[8]

2.2.4 Chemical oxygen demand

COD is defined as the amount of oxygen used while oxidizing the organic matter content of a sample with a strong chemical oxidant under acidic conditions. In COD determination, the organic matter (both biologically oxidizable like glucose and biologically inert like cellulose) is completely oxidized to CO_2 and H_2O. As it does not differentiate among them, so the COD values are greater than BOD values (which represent the amount of oxygen that bacteria need for stabilizing biologically oxidizable matter). Moreover, it does not provide any evidence of the rate at which the biologically active material would be stabilized under conditions that exist in nature.

COD test is therefore widely used for measuring the pollution strength of domestic and industrial wastes. The major advantage of COD test is that the determination is completed in 3 h, compared to the 5 days required for the BOD determination, and therefore, steps can be taken to correct errors on the day they occur. COD is used extensively in the analysis of industrial wastes. COD test is useful in indicating toxic conditions and the presence of biologically resistant organic substances. The test is widely used in the operation of treatment facilities because of the speed with which results can be obtained.

Interference and limitations

Volatile straight chain aliphatic compounds are not oxidized to any appreciable extent. This failure occurs partly because volatile organics are present in the vapor space and do not come into contact with the oxidizing agent. Straight chain aliphatic compounds are oxidized more effectively when silver sulfate (Ag_2SO_4) is added as a catalyst. The difficulties caused by the presence of halides can be overcome largely, though not completely, by complexing with mercuric sulfate ($HgSO_4$) before the refluxing procedure. A 10:1 ratio of $HgSO_4$:Cl^- is preferred. To eliminate a significant interference due to NO_2^-, add 10 mg sulfamic acid for each mg NO_2-N present in the sample volume.

Pretreatment

Samples are preserved by acidification to pH < 2 using concentrated H_2SO_4 in case immediate analysis is not done. However, the analysis is performed within the allowable period for preserved samples.

Principle

The determination of COD has been carried out using $K_2Cr_2O_7$ as the most suitable oxidizing agent, which is capable of oxidizing a wide variety of organic matters to CO_2 and H_2O. For this oxidation, the solution must be strongly acidic and at an elevated temperature. Reflux condensers are used to retard the loss of volatile organic compounds:

$$C_xH_yO_z + (4x + y - 2z/6)Cr_2O_7{}^{2-} + 4/3[4x + y - 2z]H^+$$
$$\rightarrow xCO_2 + [4x + y - 2z/3]Cr^{3+} + [16x + 7y - 8z/6]H_2O \qquad (2.7)$$

During COD determination, an excess of potassium dichromate is refluxed with a known volume of sample in an acidic condition, and the amount of excess dichromate remaining at the end of the reaction is determined by titrating it against ferrous ammonium sulfate (FAS) using ferroin as an indicator. By this titrating value, the amount of actually used dichromate during oxidation of organic matter is calculated. The reaction is represented as

$$6Fe^{2+} + Cr_2O_7{}^{2-} + 14H^+ \rightarrow 6Fe^{3+} + 2Cr^{3+} + 7H_2O \qquad (2.8)$$

The reagent blank experiment is carried out to extraneous organic matter. Certain reduced inorganic ions can be oxidized under the condition of COD test and thus can cause erroneously high results to be obtained. Cl^- causes the most serious problem because of their normally high concentration in most wastewater:

$$6Cl^- + Cr_2O_7{}^{2-} + 14H^+ \rightarrow 3Cl_2 + 2Cr^{3+} + 7H_2O \qquad (2.9)$$

Fortunately, this interference can be eliminated by the addition of $HgSO_4$ to the sample prior to the addition of other reagents. The Hg^{2+} ion combines with Cl^- ions to form poorly ionized $HgCl_2$ complex:

$$Hg^{2+} + 2Cl^- \rightarrow HgCl_2 \qquad (2.10)$$

In the presence of excess Hg^{2+} ions, the Cl^- ion is so small that it is not oxidized to any extent. $NO_2{}^-$ ions are oxidized to $NO_3{}^-$, and this interference can be overcome by the addition of sulfamic acid to the dichromate solution.

Apparatuses

a. Reflux assembly with water condensers
b. Conical flask
c. 50 mL auto burette
d. Volumetric pipette
e. 25 mL measuring cylinder

Reagents

i. *Standard potassium dichromate solution (0.25 N)*: dissolve 12.259 g $K_2Cr_2O_7$, previously dried at 13 °C for 2 h, in distilled water, add about 120 mg sulfamic acid to take care of NO_2-N, and dilute up to 1 L. Dilute 100 mL of 0.25 N $K_2Cr_2O_7$ solution to 1 L for measuring wastewater COD.

ii. *Sulfuric acid reagent*: add Ag_2SO_4 powder to concentrated H_2SO_4 at a rate of 5.5 g Ag_2SO_4/kg H_2SO_4. Let it stand for 1-2 days to dissolve Ag_2SO_4.

iii. *Ferroin indicator*: dissolve 1.485 g 1,10-phenanthroline monohydrate and 695 mg $FeSO_4 \cdot 7H_2O$ in distilled water and dilute to 100 mL. It is commercially available in the market and used as it is.

iv. *Standard FAS solution (0.1 N)*: dissolve 39.2 g of $(NH_4)_2Fe(SO_4)_2 \cdot 6H_2O$ in about 400 mL distilled water, add 20 mL of concentrated sulfuric acid, cool, and dilute to 1 L. This is standardized against 0.25 N $K_2Cr_2O_7$. Dilute 100 mL of 0.1 N FAS solution up to 1 L for measuring groundwater COD.

Actual process

25 mL of sample (or small portion diluted to 25 mL, in case of high CODs) is taken out in a 250 mL refluxing flask. Add 1 g of $HgSO_4$ and glass beads. Then 33 mL of concentrated H_2SO_4 reagent is added with care to cool the apparatus while mixing to avoid possible loss of volatile materials. 10 mL of potassium dichromate solution is added, and the flask is attached to condenser. After 2 h reflux, the mixture is cooled; condenser is washed down with distilled water. Cool to room temperature: the excess $K_2Cr_2O_7$ is titrated against FAS using ferroin indicator. Color change is from blue-green to reddish brown. Blank is also performed every time in the same manner.[10–12]

Calculation

$$COD\,(mg/L) = [(A - B) \times N \times 8 \times 1000]/mL \text{ of sample} \qquad (2.11)$$

where A is the volume of FAS used for blank (mL), B the volume of FAS used for sample (mL), N the normality of FAS, and 8 is the milliequivalent weight of oxygen.

2.2.5 Biochemical oxygen demand

BOD is simply defined as the amount of oxygen required by bacteria while stabilizing decomposable organic matter under aerobic conditions.[9,13–15] The term decomposable means the organic matter can serve as food for the bacteria and energy is derived from its oxidation. BOD is used to determine the relative oxygen requirements of wastewaters, effluents, and polluted waters. BOD gives an idea about the extent of pollution.

BOD test is essentially a bioassay procedure involving the measurement of oxygen consumed by the living organism while utilizing the organic matter present in a waste under conditions as similar as possible to those that occur in nature. Because of the limited solubility of oxygen in water, usually 9 mg/L at 20 °C, biological degradation of organic compounds under natural conditions is brought about by a diverse group of organisms that carry the oxidation essentially to completion, i.e., almost entirely to carbon dioxide and water. Therefore, it is important that a mixed group of organisms commonly called "seed" be present in the test, which is not necessary for food–based industrial wastewater as it itself survives a large community of microorganisms.

The results are obviously not reproducible; generally, the purpose of seeding is to introduce microorganism into the sample, a biological population capable of oxidizing the organic matter in the wastewater. The standard seed material is settled domestic wastewater, which has been stored for 24–36 h at 20 °C.

The determination of BOD is used to measure the purification capacity of streams and serves regulatory authorities as a means of checking the quality of effluents discharged. Information concerning BOD of waste is an important consideration in the design of treatment facilities.

Pretreatment

No preservatives are used for samples to be analyzed for BOD. The samples are only kept at 4 °C. The analysis is performed within 48 h of sample collection. The preserved samples are warmed to 27 °C before analysis. Samples are neutralized to pH 6.5–7.5 with sulfuric acid or sodium hydroxide, whenever required.

Principle

The method consists of filling with sample to overflowing in an airtight bottle of specified size and incubating it at 27 °C for 5 days. Dissolved oxygen

(DO) is measured initially and after incubation, and the BOD is computed from the difference between the initial DO and the final DO.

Apparatuses
a. Incubation (BOD) bottles: 300 mL capacity, with ground glass stoppers. Clean bottles with good detergents.
b. Incubator: thermostatically controlled at 27 °C. Exclude all light to prevent the possibility of photosynthetic production of DO (Thermo, Hi Tech).
c. Portable air pump.

Reagents
i. *Phosphate buffer solution*: dissolve 8.5 g KH_2PO_3, 21.75 g K_2HPO_4, 33.4 g $Na_2HPO_4 \cdot 7H_2O$, and 1.7 g NH_4Cl in about 500 mL distilled water and dilute up to 1 L. The pH should be 7.2 without further adjustment.
ii. *Magnesium sulfate solution*: dissolve 22.5 g $MgSO_4 \cdot 7H_2O$ in distilled water and dilute up to 1 L.
iii. *Calcium chloride solution*: dissolve 27.5 g $CaCl_2$ in distilled water and dilute up to 1 L.
iv. *Ferric chloride solution*: dissolve 0.25 g $FeCl_3 \cdot 6H_2O$ in distilled water and dilute up to 1 L.
v. *Acid solution (1 N)*: add 28 mL concentrated sulfuric acid to distilled water. Make volume up to 1 L.
vi. *Alkali solution (1 N)*: dissolve 4 g NaOH in distilled water. Dilute to up to 1 L. (These acid and alkali solutions are prepared for neutralization of alkaline or acidic waste samples.)
vii. *Sodium sulfite solution*: dissolve 1.575 g Na_2SO_3 in 1 L distilled water. This solution is not stable; prepare daily.
viii. *Glucose-glutamic acid solution*: dry reagent-glucose and reagent-grade glutamic acid at 103 °C for 1 h. Add 150 mg glucose and 150 mg glutamic acid to distilled water and dilute up to 1 L.
ix. *Ammonium chloride solution*: dissolve 1.15 g NH_4Cl in 500 mL distilled water, adjust pH to 7.2 with NaOH solution, and dilute up to 1 L. This solution contains 0.3 mg N/L.

Actual process
The BOD concentration in most wastewater samples exceeds the concentration of DO available in an air saturated sample. Therefore, it is necessary

to dilute the sample before incubation to bring the oxygen demand and supply into appropriate balance. Because bacterial growth requires nutrients such as nitrogen, phosphorous, and trace metals, these are added to the dilution water, which is buffered to ensure that the pH of the incubated sample remains in a range suitable for bacterial growth. Complete stabilization of a sample may require a period of incubation too long for practical purposes; therefore, 5 days has been accepted at 27 °C as the standard incubation period. Dilution waters are seeded further for acceptable quality of measuring their consumption of oxygen from a known organic mixture, usually glucose and glutamic acid.

The following steps are followed to perform BOD test.

Preparation of dilution water

Take desired volume of distilled water in a suitable container. 1 mL each of phosphate buffer, $MgSO_4$, $CaCl_2$, and $FeCl_3$ solutions is added per liter of distilled water. It is necessary to have a population of microorganisms capable of oxidizing the biodegradable organic matter in the sample. Here, microorganisms are already present either in domestic wastewater or surface water; seeding is not necessary, but when the sample is deficient in microorganisms, the dilution water needs seeding. The preferred and taken seed is the effluent from a biological treatment system processing the waste. 1 mL of seed is added per liter of dilution water. Air is supplied by air pump to saturate the water with DO.

Dilution technique

The BOD bottle is half-filled with dilution water. The sample portion is varying from 0.3% to 6.7%. After the addition of sample, the BOD bottle is filled with dilution water. The bottles are filled in a way that insertion of stopper displaces all air, leaving no bubbles. Two sets for each sample are prepared.

DO measurement

DO is measured by the azide modification of the Winkler method. The DO level in natural and wastewater depends on the physical, chemical, and biochemical activities in the water bodies. Oxygen is considered as poorly soluble in water. Its solubility is related to pressure and temperature. In freshwater, DO reaches 14.6 mg/L at 0 °C and approximately 9.1, 8.3, and 7.0 mg/L at 20, 25, and 35 °C, respectively, and 1 atm pressure. At temperatures of 20 and 30 °C, the level of saturated DO is 9.0–7.0 mg/L. Low oxygen in water can kill fish and other organisms present in water. For living organism, about 4 mg/L of minimum DO should be in water. The oxygen–depleting substances reduce the available DO. During summer months, the rate of biological oxidation is highly increased. Unfortunately, the DO concentration is at its minimum due to higher temperature.

The azide modification of the Winkler method.

Principle

It is a titrimetric procedure based on the oxidizing property of DO. Oxygen present in the sample oxidizes the divalent manganese to its higher valency, which precipitates as brown hydrated oxides after the addition of NaOH and KI. Upon acidification, manganese reverts to divalent state and liberates iodine from KI equivalent to DO content in the sample. The liberated iodine is titrated against $Na_2S_2O_3$ using starch as an indicator.

The series of reactions can be summarized below:

$$MnSO_4 + 2NaOH \rightarrow Mn(OH)_2 \downarrow (\text{white ppts}) + Na_2SO_4 \qquad (2.12)$$

If no oxygen is present, a pure white precipitate of $Mn(OH)_2$ forms when $MnSO_4$ and alkali-iodide (NaOH + KI) are added to the sample. If oxygen is present, the divalent Mn(II) is oxidized to higher–valency Mn (IV) and precipitated as a brown hydrated oxide (MnO_2):

$$Mn(OH)_2 + \frac{1}{2}O_2 \rightarrow MnO_2 + H_2O \qquad (2.13)$$

The oxidation of Mn(II) to Mn(IV) is sometimes called fixation of the oxygen. Under acid condition, MnO_2 reverts to divalent state by oxidizing KL to produce I_2:

$$MnO_2 + 2I^- + 4H^+ \rightarrow Mn^{2+} + I_2 + 2H_2O \qquad (2.14)$$

The liberated I_2 is titrated against standard solution of sodium thiosulfate ($Na_2S_2O_3$).

Reagents

i. *Manganese sulfate solution ($MnSO_4 \cdot H_2O$)*: dissolve 364 g of monohydrate manganese sulfate in distilled water, filter it if necessary, and dilute up to 1 L. This solution should not give color with starch solution when added to an acidified solution of potassium iodide (KI).

ii. *Alkaline-iodide-azide solution*: dissolve 500 g NaOH and 135 g NaI and dilute to 950 mL. Add 10 g of sodium azide (NaN_3) dissolved in 40 mL distilled water. Cool the solution, and make volume up to 1 L. This solution should not give color with starch solution when diluted and acidified.

iii. *Starch indicator:* take 5 g of soluble starch to approximately 800 mL of boiling water, with stirring. Dilute up to 1 L, boil for a few minutes, and leave overnight. Use clear supernatant. Preserve by adding a few drops of toluene or formalin. Store it in a bottle with glass stopper.

iv. *Standard sodium thiosulfate solution (0.025 N):* dissolve directly 6.025 g $Na_2S_2O_3 \cdot 5H_2O$ in a 1 L previously boiled and then cooled distilled water, and add 1.5 mL 6 N NaOH or 0.4 g solid NaOH per 1 L. Store in a brown bottle.

v. *Standard potassium dichromate solution (0.25 N):* dry $K_2Cr_2O_7$ at 103 °C for 2 h and take accurately 12.259 g, and dilute to up to 1 L (used for standardizing thio solution).

Actual process

The following steps are followed for DO measurement.

Add 1 mL of $MnSO_4$ followed by alkali–iodide–azide solution. Place the stopper carefully to exclude air bubbles and mix by inverting the bottle repeatedly for a few minutes. (An equivalent amount of 2 mL of the contents will come out of the bottle after placing the stopper.) The precipitation obtained represents the presence of oxygen. Allow the precipitate to settle leaving about 150 mL of clear supernatant. Carefully remove the stopper and immediately add 1 mL of concentrated sulfuric acid. Close the bottle and mix with gentle inversion until the precipitate completely dissolves. Titrate 200 mL contents of the bottle with sodium thiosulfate solution using starch as an indicator. The blue color turns to colorless.

Calculation

$$DO \text{ in } mg/L = [mL \text{ of titrant} \times \text{Normality} \times 8 \times 1000]/[V_2(V_1 - V)/V_1]$$

(2.15)

where V_1 is the volume of BOD bottle, V_2 is the volume of contents titrated (mL), and V is the volume of $MnSO_4$ and iodide azide added, i.e., $1 + 1 = 2$ mL.

Thus, the first initial DO is determined for one set of BOD. The second set of bottles is kept for incubation in the incubator and the temperature is maintained at 27 °C for 5 days. The DO of these bottles is measured on completion on the fifth day.

Calculation for BOD$_5$
BOD is calculated below:

$$BOD_5 \text{ in } mg/L = [(D_1 - D_2) - (B_1 - B_2)f]/p \qquad (2.16)$$

where D_1 is the DO of diluted sample immediately after preparation (mg/L), D_2 the DO of diluted sample after 5-day incubation (mg/L), P the decimal volumetric fraction of sample used, B_1 the DO of seed control before incubation (mg/L), B_2 the DO of seed control after incubation (mg/L), and f is the ratio of seed in diluted sample to seed in seed control $= (1 - P)$.

2.2.6 Hardness
Hard waters are generally considered to be those waters that require considerable amounts of soap to produce foam or lather and that also produce scale in water pipes, heaters, boilers, and other units in which the temperature of water is increased materially. Originally, water hardness is considered as a measure of the capacity of water to precipitate soap. Hardness is caused by multivalent metallic cations from sedimentary rocks, seepage, and runoff from soils. Such ions are capable of reacting with soap to form precipitates and with certain anions present in the water to form scale. Calcium and magnesium, the two principal ions, are present in many sedimentary rocks, the most common being limestone and chalk. They are also present in a wide variety of industrial products and are common constituents of food. As mentioned above, a minor contribution to the total hardness of water is also made by other polyvalent ions, e.g., aluminum, barium, iron, manganese, strontium, and zinc. The hard water may have some adverse effects on human beings, i.e., cancers and cardiovascular diseases.

In conformity with the current practice, total hardness is defined as the sum of calcium and magnesium concentrations, both expressed as calcium carbonate (in mg/L). It is further defined in two types: carbonate hardness and noncarbonate hardness. Carbonate hardness is essentially important since it leads to scaling. Noncarbonate hardness is formerly called permanent hardness, because it cannot be removed by boiling. Noncarbonate hardness cations are associated with sulfate, chloride, and nitrate ions of calcium and magnesium.

The hardness is divided into three types according to its analysis: (i) total hardness, (ii) calcium hardness, and (iii) magnesium hardness.

Principle

This method involves the use of solution of ethylenediaminetetraacetic acid (EDTA) or its sodium salt as the titrating agent. In alkaline conditions, EDTA or its sodium salt (Na_2EDTA) reacts with calcium and magnesium to form a soluble chelated complex. This chelated complex gives color changes in the presence of indicator. EDTA reacts with calcium and magnesium, but it combines with only calcium at higher pH (above 12). So, calcium or calcium hardness is determined directly by EDTA, when higher pH (above 12) is maintained using 1 N NaOH during titration and the indicator used for titration is murexide. All of calcium and magnesium are complexed by EDTA at a pH of 10. This pH is maintained by ammonia buffer during titration process. Eriochrome black T (EBT) is used instead of murexide.

Total hardness
Apparatuses
a. 250 mL measuring cylinder
b. 250 mL conical flask
c. Burette

Reagents
  i. *Ammonia buffer solution*: dissolve 16.9 g NH_4Cl in 143 mL concentrated NH_4OH. Add 1.25 g EDTA salt (commercially available) and dilute to 250 mL with distilled water. Store in a borosilicate glass bottle. Place a stopper tightly to avoid ammonia loss and CO_2 intake. Store this solution for a maximum of 1 month. Then after, discard it and prepare fresh solution to determine the total hardness.
 ii. *EBT*: (Sodium salt of 1-(1-hydroxy-2-naphthylazo)5-nitro-2-naphthol-4-sulfonic acid, No. 203 in the color index.) Dissolve 0.5 g dye in 100 mL of triethanolamine.
iii. *0.1 M EDTA solution*: weight 37.23 g analytic reagent-grade disodium ethylenediaminetetraacetate dihydrate, and dilute up to 1 L with distilled water. Standardize against $CaCO_3$ solution.

Titration precautions
Conduct titration at room temperature. Complete titration within 5 min to avoid $CaCO_3$ precipitation.

Actual process

100 mL sample volume is taken out in a conical flask. Buffer solution is added until pH 10. Two drops of indicator are added, and the mixture is titrated against 0.1 M EDTA solution until the wine's red color changes to blue. Note the reading "A." Run a reagent blank with distilled water. Note the reading "B." Calculate the volume of EDTA required by sample $= C = (A - B)$ mL.[15–19]

Calculation

$$\text{Total hardness (mg/L)} = (C \times D \times 1000)/100 \, \text{mL (sample taken)} \quad (2.17)$$

where C is the mL titration for sample and D is the mg $CaCO_3$ equivalent to 1.00 mL EDTA titrant.

Calcium hardness

Apparatuses

a. 250 mL measuring cylinder
b. 250 mL conical flask
c. Burette

Reagent

i. *0.1 M EDTA solution*: weight 37.23 g analytic reagent-grade disodium ethylenediaminetetraacetate dihydrate, and dilute up to 1 L with distilled water.
ii. *Murexide*: prepare a ground mixture of 200 mg of murexide with 100 g of solid NaCl.
iii. *N sodium hydroxide*: dissolve 80 g NaOH and dilute to 1 L.

Actual process

Take about 100 mL sample in a conical flask. Add 1 mL NaOH to raise pH to 12.0 and a pinch of murexide. Titrate immediately with EDTA until pink color changes to purple. Note the volume of EDTA required (A_1). Run a reagent blank with distilled water. Note the mL of EDTA required (B_1), and keep it aside to compare end points of sample titrations. Calculate the volume of EDTA required by sample $= C_1 = (A_1 - B_1)$ mL.

Calculation

$$\text{Calcium Hardness (as mg Ca/L)} = (C_1 \times D \times 400.8)/\text{mL sample}$$

(2.18)

where C_1 is the mL titration for sample and D is the mg $CaCO_3$ equivalent to 1.00 mL EDTA titrant.

Magnesium hardness

Magnesium hardness can be determined by calculating the difference between the total hardness and the calcium hardness of the sample.[16–19]

$$\text{Mg hardness} = \text{Total hardness} - \text{Calcium hardness}$$

(2.19)

Or it can be calculated by

$$\text{Mg hardness as Mg} = [(C - C_1) \times D \times 243]/\text{mL sample}$$

(2.20)

where C is the mL titration for total hardness, C_1 the mL titration for calcium hardness, and D is the mg $CaCO_3$ equivalent to 1.0 mL EDTA titrant.

2.2.7 Oil and grease

Oil and grease are useful to determine treatment plant efficiencies and to overcome the difficulties during treatment. The presence of oil and grease in trade effluents is quite common. The sources for oil and grease are natural raw materials used in the process and/or from the lubricants applied for machineries. The oil and grease fraction is reported to contain hydrocarbons, lipids, fatty acids, soaps, fats and waxes, and oils. It should be stated that results are obtained by solvent extraction method that indicates not only the oil and grease content but also the quantity of all extractable matter by the solvent. It does not indicate the quantity of volatile oil and grease. However, the results are useful for practical considerations; when other constituents of organic matter are more, the results should be reviewed and interpreted carefully.

Principle

Oil, grease, and other extractable matters are dissolved in a suitable solvent and extracted from the aqueous phase. The solvent layer is then evaporated and the residue is weighted as oil and grease.

Actual process

Place 100 mL of the well-mixed sample in a beaker. Add 5 mL $MgSO_4$ solution (1%) to the sample and while stirring the small amount of milk of lime (2%) until flocculation occurs. Continue stirring for 2 min; allow the precipitates to settle for 5 min. When the precipitates have completely settled, siphon off the top layer leaving about 1.5 cm clear layer above the precipitate level. Dissolve the precipitates in dilute HCl $(1+3)$ and transfer the liquid to the separating funnel, wash the beaker with 50 mL of petroleum ether, and add this to the separating funnel. Shake the funnel continuously and gently for 1 min. Draw off the aqueous layer into another separating funnel and again extract with 50 mL of petroleum ether. Combine the extracts in a beaker and pass them through the Whatman filter paper No. 1 containing sodium sulfate in its cone and moistened with the solvent, collected into evaporating dish, and keep it on a water bath. Dry the outside, cool, and weigh (W_2). The difference in weight is the amount of oil present in the aliquot of the sample. Weigh the empty evaporating dish (W_1).[5,6]

Calculation

$$\text{Oil and grease (mg/L)} = [(W_2 - W_1) \times 1000]/\text{Volume of the sample taken (mL)}$$

$$(2.21)$$

2.2.8 Chloride

Natural water, industrial wastewater, and even some surface water contain chlorides in varying amounts. Chloride, in the form of chloride ion, is one of the major inorganic anions in water and wastewater. In potable water, the salty taste produced by chloride concentrations is variable and dependent on the chemical composition of water. Some waters containing 250 mg/L may have a detectable salty taste if the cation is sodium. On the other hand, the typical salty taste may be absent in waters containing as much as 1000 mg/L when predominant cations are calcium and magnesium. The chloride concentration is higher in wastewater than in raw water because sodium chloride is a common article of diet and remains unchanged as it passes through the digestive system. Along the sea coast, chloride may be present in higher concentrations because of the leakage of saltwater into the sewage system. It also may be increased by industrial process. A chloride content may harm metallic pipes and structures, as well as growing plants.

Principle

Chloride ions in a neutral or faintly alkaline solution can be estimated by titration with a standard solution of $AgNO_3$ using K_2CrO_4 as an indicator (Mohr's method). The pH must be in the range 7-8 because Ag^+ ions are precipitated as AgOH at higher pH and CrO_4^{2-} is converted to $Cr_2O_7^{2-}$ at lower pH:

$$Ag^+ + OH^- \rightarrow AgOH \left(K_{sp} = 2.3 \times 10^{-8}\right) \qquad (2.22)$$

$$2CrO_4^{2-} + 2H^+ \rightarrow 2HCrO_4^- \rightarrow Cr_2O_7^{2-} + H_2O \qquad (2.23)$$

$HCrO_4^-$ being a weak acid, CrO_4^{2-} ion concentration is decreased necessitating higher concentration of Ag^+ for the solubility product of Ag_2CrO_4 to be exceeded, thus leading to higher results. The required pH range can be achieved easily by adding a pinch of pure $CaCO_3$ to the pink or red solution obtained at the end point of the methyl orange alkalinity determination. Excess $CaCO_3$ being insoluble does not interfere:

$$2H^+ + CaCO_3 \rightarrow Ca^{2+} + CO_2 + H_2O \qquad (2.24)$$

As $AgNO_3$ solution is added from the burette to the chloride ion sample containing CrO_4^{2-}, Ag^+ react with both Cl^- and CrO_4^{2-}, forming the respective precipitates:

$$Ag^+ + Cl^- \xrightarrow[\text{White ppts}]{} AgCl \left(K_{sp} = 3 \times 10^{-10}\right) \qquad (2.25)$$

$$2Ag^+ + CrO_4^{2-} \rightarrow Ag_2CrO_4 \left(K_{sp} = 5 \times 10^{-10}\right) \qquad (2.26)$$

But the red color formed by the addition of each drop of $AgNO_3$ disappears because of the large concentration of Cl^- ion in the solution:

$$Ag_2CrO_4 + 2Cl^- \rightarrow 2AgCl + CrO_4^{2-} \qquad (2.27)$$

As the concentration of Cl^- ions decrease, the red color disappears more slowly and when all the chloride has been precipitated, a faint reddish to pinkish tinge persists in the white precipitate even after brisk shaking. If the wastewater sample pH is acidic in nature (below 6.5), it is necessary to neutralize the sample by the addition of $NaHCO_3$ until completion of effervescing of CO_3^{2-} of the sample; otherwise, sharp color change of the water sample near the end point cannot obtained.

Reagents

i. *Potassium chromate*: dissolve 50 g K_2CrO_4 in a little distilled water. Add $AgNO_3$ solution until a definite red precipitate is formed. Let it stand for 12 h, filter, dilute to 1 L.

ii. *Standard silver nitrate solution (0.02 N)*: take 4.79 g $AgNO_3$ and dilute up to 1 L with distilled water. Standardize it against 0.02 N NaCl (1648 dry mg dilute to 1 L). For higher concentration of chloride in wastewater samples, 1 mL of $AgNO_3 = 1$ mg Cl^- std solution (6.754 g dilute to 1 L) is used, which is 0.0282 N.

Actual process

100 mL of sample is taken out in 250 mL conical flask. The titration is to be done in the pH range 7.0-10. Hence, adjust the pH by 0.1 N H_2SO_4 or 0.25 N NaOH wherever required. Then potassium chromate is added. Yellow color appears. The mixture is titrated against standard silver nitrate titrant. Silver chloride is precipitated quantitatively. In the end, silver chromate red buff precipitates out. Established reagent blank value by titration method is outlined as above.[20,21]

Calculation

$$mg\, Cl^-/L = (A - B) \times N \times 35.5 \times 1000/100\ mL\ (sample\ taken)\quad (2.28)$$

where A is the volume of standard $AgNO_3$ used for sample (mL), B is the volume of standard $AgNO_3$ used for blank (mL), and N is the normality of $AgNO_3$.

2.2.9 Phenol

Phenols are the most important groups of aromatic compounds. It is commonly known as carbolic acid. It has been used widely as a germicide and disinfectants. Phenol can be treated biologically up to 700 mg/L aerobically and up to 200 mg/L anaerobically. Phenolic compounds are sometimes found in surface natural waters and industrial sources. The phenols in the water environment can arise from natural substance degradation, industrial activities, and agricultural practices. Chlorinated phenols may be life-threatening to humans even at low concentration. Their presence gives a disagreeable smell and taste even at low ppm concentrations. The US Environmental Protection Agency includes in the Federal Register list eleven substituted phenol retained hazardous for human health and assigns them a maximum admissible concentration range of 60–400 mg/L in relation to their toxicity degree. The presence of varied types of inorganic substances in effluent and polluted waters interferes with the colorimetric procedure, and samples should be subjected to preliminary treatment, distillation, and extraction. The phenols are separated from other nonvolatile impurities by distillation. As the rate of volatilization of phenol is gradual, the volume of the distillate must be equal to that of sample being distilled.

Phenol is defined as hydroxyl derivatives of benzene, and its condensed nuclei may occur in domestic and industrial wastewater. Chlorination of such waters may produce odoriferous and objectionable-tasting chlorophenols, which may include o-chlorophenol, p-chlorophenol, 2,6-dichlorophenol, and 2,4-dichlorophenol.

Principle

The steam distillable phenols react with 4-aminoantipyrine at a pH of 7.9 in the presence of potassium ferricyanide to form a colored antipyrine dye. The dye is extracted from aqueous solution with chloroform, and the intensity is measured at 460 nm. This method is applicable in the concentration range of 1–250 µg/L with a sensitivity of µg/L.

Pretreatment

Samples were preserved in case of delayed analysis by acidifying them with 2 mL of concentrated H_2SO_4/L. and stored at 4 °C. However, the analysis was performed within 28 days after collection.

Interferences and limitations

Interferences such as phenol-decomposing bacteria, oxidizing and reducing substances, and alkaline pH values are dealt with by acidification. The 4-aminoantipyrine colorimetric method determines phenols: ortho- and meta-substituted phenols, under proper pH conditions, and para-substituted phenols in which the substitution is a carboxyl, halogen, methoxy, or sulfonic acid group. It does not determine those para-substituted phenols where the substitution is an alkyl, aryl, nitro, benzyl, nitroso, or aldehyde group. Because the relative amounts of various phenolic compounds in a given sample are unpredictable, it is not possible to provide universal standard containing a mixture of phenols. For this reason, phenol (C_6H_5OH)

itself has been selected as a standard for colorimetric procedure and any color produced by the reaction of other phenolic compounds is reported as phenol.

Apparatuses and equipment

a. Distillation assembly: all glass consisting of 1 L Pyrex distilling apparatus with Graham condenser
b. Spectrophotometer (Spectroquant NOVA 60)
c. pH meter (Equiptronics; Model: EQ-614 A)

Reagents and standards

i. *Phosphoric acid*: dilute 10 mL 5% H_3PO_4 to 100 mL with distilled water.
ii. *Methyl orange*: dissolve 0.5 g methyl orange in 1 L distilled water.
iii. *1 N sulfuric acid*: dilute 28 mL of concentrated H_2SO_4 to 1 L with distilled water.
iv. *2.5 N sodium hydroxide*: dissolve 10 g NaOH in 100 mL distilled water.
v. *Stock phenol solution*: dissolve 1 g phenol in freshly boiled and cooled distilled water and dilute to 1 L. Standardize the stock phenol solution. 1 mL = 1 mg phenol.
vi. *Intermediate phenol solution*: take 10 mL or appropriate volume of stock phenol solution in 1 L volumetric flask and dilute to the mark with freshly boiled and cooled distilled water to get 1 mL = 10 µg phenol.
vii. *Standard phenol solution*: dilute 50 mL intermediate phenol solution to 500 mL with freshly boiled and cooled distilled water. This solution should be prepared within 2 h of use. 1 mL = 1 mg phenol.
viii. *0.1 N bromate bromide solution*: dissolve 2.784 g anhydrous $KBrO_3$ in water, add 10 g KBr crystals, dissolve, and dilute to 1 L with distilled water.
ix. *0.025 N standard sodium thiosulfate*: dissolve 6.205 g $Na_2S_2O_3 \cdot H_2O$ in distilled water and dilute to 1 L with distilled water.
x. *0.5 N ammonium hydroxide*: dilute 35 mL fresh concentrated NH_4OH to 1 L with distilled water.
xi. *Phosphate buffer solution*: dissolve 104.50 g K_2HPO_4 and 72.3 g KH_2PO_4 in distilled water and dilute to 1 L; the pH of this solution should be 6.8.
xii. *4-Aminoantipyrine solution*: dissolve 2.0 g 4-aminoantipyrine in distilled water and dilute to 100 mL. Prepare this solution daily.
xiii. *Potassium ferricyanide solution*: dissolve 8.0 g $K_3Fe(CN)_6$ in water and dilute up to 100 mL using distilled water. Filter if necessary, and store in a brown glass bottle.

Actual process

1. *Distillation*

   A. Measure 500 mL sample into a beaker, add 50 mL phenol-free distilled water, and lower the pH to 4.0 with H_3PO_4 solution using methyl orange as an indicator. Add 5 mL $CuSO_4$ solution. Transfer to distillation flask and collect 500 mL distillate using measuring cylinder as receiver. If the distillate is turbid, repeat the same procedure as above. Omit the addition of H_3PO_4 and $CuSO_4$ if the preserved sample is used.

   B. Take 500 mL of the original sample. Make it acidic with 1 N H_2SO_4 using methyl orange as an indicator. Transfer to a separating funnel and add 150 g NaCl. Shake with five increments of chloroform, using 40 mL in the first increment and 25 mL in each of the following increments. Transfer the chloroform layer to another separatory funnel and shake with three successive increments of 2.5 N NaOH solution using 4.0 mL in the first increment and 3.0 mL in each of the next two increments. Combine the alkaline extracts. Heat on water bath until the chloroform has been removed. Cool and dilute to 500 mL with distilled water and proceed to distillation as in (A).

2. *Extraction and color development*

Take 500 mL of the distillate or a suitable portion containing more than 50 mg phenol and dilute to 500 mL in 1 L beaker. Take 500 mL distilled water blank and a series of 500 mL phenol standards containing 5, 10, 20, 30, 40, and 50 µg phenol, in respective beakers. Add 12 mL 0.5 N NH_4OH solution and adjust the pH of each to 7.9 ± 0.1 with 10 mL phosphate buffer. Transfer to 1 L separating funnel, add 3.0 mL 4-aminoantipyrine solution in each separatory funnel, mix well, and add 3.0 mL potassium ferricyanide; again, mix well and let the color develop for 15 min. Add 25 mL chloroform in each separatory funnel and shake at least 10 times, and let the $CHCl_3$ settle again. Filter each $CHCl_3$ extract through filter paper containing 5 g layer of anhydrous Na_2SO_4. Collect dried extract clean cells, and measure the absorbance of sample and standard against the blank at 460 nm. Plot absorbance against mg phenol concentration, and draw a calibration curve. Estimate sample phenol content from photometric reading by using a calibration curve.[5,6]

Calculations

Use a calibration curve:

$$\text{µg/L, phenol} = [(A / B) \times 1000] \qquad (2.29)$$

where A is the µg phenol in sample.

2.2.10 Total dissolved solids

Dissolved solids are the major concern in water samples; therefore, the TDS determinations and the specific conductance measurements are of great interest. TDS determinations are ordinarily of little value in the analysis of polluted water because they are difficult to interpret with any degree of accuracy. Intermittent discharge of highly mineralized waste may bring changes in density; for this purpose, total dissolve solid test can be used to good advantage to detect such change.

Industrial wastes include such a wide variety of materials that analysis for exploratory purpose should include all determinations that can possibly provide significant information. For this reason, all solid tests that commonly applied domestic wastewater are important. Many industrial wastes contain unusual amounts of dissolved inorganic salts, and their presence is easily detected by the total solid test. Their concentration and nature are factors in determining the susceptibility of waste to anaerobic treatment. Highly mineralized water with a considerable calcium, magnesium chloride, and/or sulfate content may be hygroscopic and require prolonged drying proper, desiccation, and rapid weighing. Samples high in bicarbonate require careful and possibly prolonged drying at 180 °C to ensure complete conversion of bicarbonate to carbonate.

Apparatuses
a. Evaporating dishes: dishes of 100 mL capacity made up of porcelain
b. Desiccator
c. Hot air oven (Labtech by DR Scientific) (range: 0–300 °C)
d. Analytic balance (Contact; Model: CB-50)

Actual process
An evaporating dish of 100 mL capacity is taken. It is cleaned, dried at 103–105 °C in oven for 1 h, and then cooled in a desiccator. It is weighed before use and the initial weight (W_1) is noted. Appropriate volume of sample is filtered through Whatman filter paper No. 1 so that the filtrate should not have any turbidity. Heat it on a water bath. After evaporation, heat it at 103 °C for 1 h in an oven. After all water is evaporated, the evaporating dish is cooled in a desiccator and the final weight (W_2) is taken.[22,23]

Calculation

$$\text{TDS}\,(\text{in mg/L}) = [(W_2 - W_1)/V] \times 1000 \qquad (2.30)$$

where W_1 is the initial weight of the evaporating dish (g), W_2 the final weight of the evaporating dish (g), and V is the volume of the sample taken (mL).

2.2.11 Total alkalinity

Alkalinity is a measure of the capacity of water to neutralize acids. Alkalinity of water is due primarily to the presence of bicarbonate, carbonate, and hydroxide ions. Salts of weak acids, such as borates, silicates, and phosphates, may also contribute. Salts of certain organic acids may contribute to alkalinity in polluted or anaerobic water, but their contribution usually is negligible. Bicarbonate is the major form of alkalinity. The alkalinity of a sample is the measure of its capacity to neutralize acids. Knowledge of the kind of alkalinity present in water and their magnitude is important. The alkalinity of water refers to its acid–neutralizing capacity. It is the sum of all titratable bases. Highly alkaline waters, above pH 7.0, can create a bitter taste and a slippery feel and also cause drying of the skin. Alkalinity is important for fish and aquatic life because it protects or buffers against rapid pH changes and makes water less vulnerable to acid rain, protecting the major source of human consumption.

Principle

The alkalinity of a water sample may be determined volumetrically by titrating it with a standard acid to an arbitrary pH using an indicator. When the pH of the sample is above 8.3, titration is first carried out using phenolphthalein. At the end point, when the indicator changes from pink to colorless, the pH is lowered to about 8.3.

In natural water, most H^+ of the alkalinity is due to CO_2. The free CO_2 dissolves in water to form carbonic acid (H_2CO_3), which further dissociates into H^+ and HCO_3^-. Thus, HCO_3^- formed further dissociates into H^+ and CO_3^-:

$$CO_2 + H_2O \rightarrow H_2CO_3\,(\text{dissolved } CO_2 \text{ and carbonic acid}) \qquad (2.31)$$

$$H_2CO_3 \rightarrow\ + HCO_3^-\,(\text{bicarbonate}) \qquad (2.32)$$

$$HCO_3^- \rightarrow H^+ + CO_3^-\,(\text{carbonate}) \qquad (2.33)$$

Thus, alkalinity meq/L$=[HCO_3^-]+[CO_3^-]+[OH^-]-[H^+]$ where the quantities in parentheses are concentrations in meq/L or mg/L as $CaCO_3$.

Hydroxyl ions present in a sample as a result of dissociation or hydrolysis of solutes react with additions of standard acid. Alkalinity thus depends on the end point pH used. Titrate at room temperature with a properly calibrated pH meter or electrically operated titrator or use color indicators.

Apparatuses
a. 250 mL measuring cylinder
b. 250 mL conical flask
c. Burette

Reagents
i. *Sulfuric acid (0.1 N)*: take 3 mL concentrated H_2SO_4 and dilute to 1 L with distilled water.
ii. *Standardize against Na_2CO_3 solution (0.05 N)*: take 2.5 ± 0.2 g (dry) and dilute up to 1 L with distilled water. Keep no longer than 1 week.
iii. *Phenolphthalein*: Dissolve 0.5 g phenolphthalein in 500 mL 95% ethyl alcohol, and add 500 mL distilled water. It is commercially available in the market and used as it is. (1% phenolphthalein solution used in this experiment is purchased from Merck.)
iv. *Methyl orange*: dissolve 0.5 g methyl orange powder in distilled water and dilute up to 1 L. It is commercially available in the market and used as it is. (0.04% Methyl orange solution used in this experiment is purchased from Merck.)

Actual process
100 mL water sample is taken out in a 250 mL conical flask; add 2–3 drops of phenolphthalein; if the color changes to pink, titrate it with 0.1 N H_2SO_4 until color disappears. End point will be pink to colorless. Note the reading "*P*." Next, add 2–3 drops methyl orange to the same flask, and continue titration until color changes from yellow to orange. Note "*M*" from "*P*."[24]

$$\text{Total reading } T = P + M$$

Calculation

$$\text{Total alkalinity}\,(\text{mg/L as CaCO}_3) = (A \times N \times 50 \times 1000)/\text{mL of sample}$$
(2.34)

where A is the burette reading ($= T$) and N is the normality of acid. (For phenolphthalein, alkalinity as $CaCO_3$ $A = P$, for methyl orange, alkalinity as $CaCO_3$ $A = M$, and for total alkalinity as $CaCO_3$ $A = T$.)

2.2.12 Fluoride

Fluorine (F) is a naturally occurring element found in our air, soil, and water. Fluoride is the most electronegative element and most abundant element in the Earth's crust. Fluorides can be found in relative abundance in our daily lives, most commonly in our drinking water either added artificially as a dental aid or naturally occurring in some areas of the country. It occurs as fluorspar or cryolite. The concentration of fluoride in natural water mainly depends on the solubility of the fluoride-bearing rocks with which water is in contact. The concentration of fluoride in seawater is 0.8–1.4 ppm. In some countries, the occurrence of fluoride in surface waters is widespread. Fluoride may occur naturally in water or it may be added in a controlled manner.

Exceptionally, fluorides are used as insecticides and in many industrial processes. Fluoride is highly toxic when ingested in sufficient quantity. Fluoride with lower concentration at an average of 1 mg/L is regarded as an essential constituent of drinking water because of its role in the prevention of dental carries. Exposure to excessive free fluorides can be fatal to humans and all life forms. An appropriate concentration of fluoride in the drinking water is required to prevent dental cavities, but long-term ingestion of water that contains excess concentration of fluoride (>1.5 mg/L) causes various diseases like osteoporosis, arthritis, brittle bones, cancer, infertility, brain damage, Alzheimer syndrome, thyroid disorder, bone disease, and mottling of teeth.

Principle

The fluoride electrode is an ion–selective sensor. The key element in fluoride electrode is the laser-type doped lanthanum fluoride crystal across which a potential is established by fluoride solutions of different concentrations. The crystal contacts the sample solution at one face and an internal reference solution at the other. The cell may be represented by

$$Ag/AgCl, Cl^- \, (0.3\,M), \, F^- \, (0.001\,M)/LaF_3/test\,solution/reference\,electrode$$

The fluoride electrode can be used with a standard calomel reference electrode and almost any modern pH meter having an expanded millivolt scale. The F^- ion activities at the two faces of the crystal are different, and so the potential is established, and since the conditions at the internal face are constant, the resultant potential is proportional to the F^- ion activity of the solution. The measured potential corresponding to the level of F^- ions in the solution is described by the Nernst equation:

$$E_{means} = \text{Const.} + (RT/nF)\log a_{F^-} - E_{SCE} \qquad (2.35)$$

$$E_{means} = 0.058\log[F^-] + \text{Const.} \quad \text{or} \quad E = E_0 - s\log A \qquad (2.36)$$

where E is the measured electrode potential, E_0 the reference potential (constant), A the fluoride level in solution, and S is the electrode slope.

The electrode responds only to the free ions.

Reagents

i. *Stock fluoride solution*: dissolve 221.0 mg anhydrous sodium fluoride, NaF, in distilled water and dilute to 1 L; 1 mL $= 100$ µg (or 0.1 mg) F^-. The solution is commercially available in the market and used as it is.

ii. *Standard fluoride solution*: dilute 100 mL stock fluoride solution to 1 L with distilled water, 1 mL $= 10.0$ µg F^-. Prepare a series of standard fluoride solution for 5 and 1 mg/L.

iii. *Fluoride buffer [total ionic strength adjustment buffer (TISAB) II solution]*: dissolve 57 mg glacial acetic acid, 58 g NaCl, and 4 g CDTA (1,2-cyclohexylenedinitrilotetraacetic acid, also known as cyclohexa-nediaminetetraaceticacid) in 500 mL distilled water contained in a large beaker. Stir well. Place the beaker in a water bath for cooling. Immerse a calibrated pH electrode into the solution, and slowly add approximately 5 M NaOH until the pH is between 5.0 and 5.5. Cool to room tem-perature. Pour into a 1 L flask and dilute to mark with distilled water. The solution is commercially available in the market and used as it is. (Stock fluoride solution and TISAB II solution used in this experiment are purchased from Thermo Electron Corporation.)

The buffering procedure is necessary because OH^- ions having the same charge and similar size to the F^- ion act as an interference with the LaF$_3$ electrode. CDTA will complex any polyvalent ions (such as iron and alumi-num), which may interact with fluoride. CDTAs also provide a constant background ionic strength.

Actual process

The following steps are followed for fluoride measurement through the instrument by Thermo Electron Corporation.

Set up the instrument as per manual. Here, the F^- ion electrode and reference (calomel) electrode are combined in one electrode. As the instrument can also work as a pH meter, select F^- option. Calibrate the instrument using 10, 5, and 1 mg/L standards and adjust slope at 54-60. The instrument is thus ready for sample measurement. In 20 mL sample, add 2 mL buffer solution (TISAB II). Let electrode remain in the solution to a depth of about 1 in. for 3 min (or until reading is constant). The instrument gives direct reading in mg/L. Withdraw electrode, rinse with distilled water, and blot-dry between readings. Repeat measurements with samples. Electrode is stored in distilled water or in standardizing solution.[5]

2.2.13 Sulfate

In natural water, the sulfate ion is present in appreciable amounts. When it is present in water in excess amount, it shows cathartic effect on human beings. It has the tendency to form hard scale in boilers and heat exchangers in public and industrial water supplies.

Sulfate is one of the least toxic anions. However, catharsis, dehydration, and gastrointestinal irritation are observed at higher concentrations. Waters with 300-400 mg/L sulfate have a bitter taste and those with 1000 mg/L or more of sulfate may cause intestinal disorder. Sulfates are indirectly responsible for two serious problems during handling and treatment of wastewater, i.e., odor and sewer corrosion problems resulting from the reduction of sulfates to H_2S by bacteria under anaerobic conditions. Bacteria capable of oxidizing hydrogen sulfide to sulfuric acid are always present in the domestic wastewater, and this leads to the serious problems of sewer corrosion, and H_2S itself has a bad odor. The contents of sulfate present in natural water have great importance for suitability of public and industrial water supplies. High sulfate may cause dehydration and diarrhea in human beings. Kids are often more sensitive to sulfate than adults. Also, high levels of sulfate may cause severe and chronic diarrhea and, in some cases, death in animals.

Principle

Sulfate can be estimated by gravimetric method or by turbidimetric method. Sulfate ions are precipitated in a hydrochloric acid medium as barium sulfate ($BaSO_4$) by the addition of barium chloride ($BaCl_2$):

$$BaCl_2(excess) + SO_4^{2-} \rightarrow BaSO_4 \downarrow (precipitate) \qquad (2.37)$$

Light absorbance of the $BaSO_4$ suspension is measured by spectrophotometer, and the SO_4^{2-} concentration is derived from the graph. In potable waters, there are no ions other than SO_4^{2-} that will form insoluble compounds with barium under strong acidic conditions. The minimum detectable concentration from this process is 1 mg SO_4^{2-}/L.

Interference

Color or suspended matter in large amounts will interfere. The suspended matter may be removed by filtration. Silica in excess of 500 mg/L will interfere. Water containing large quantities of organic material may not be possible to precipitate $BaSO_4$ satisfactorily.

Reagents

i. *BaCl₂ solution*: dissolve 100 g $BaCl_2 \cdot 2H_2O$ in 1 L distilled water. Filter through a membrane filter before use. 1 mL of this reagent is capable of precipitating approximately 40 mg SO_4^{2-}.

ii. *Hydrochloric acid (50%, v/v)*: add 50 mL distilled water to 50 mL of concentrated HCl.

iii. *Sulfuric acid (0.01 N)*: dilute 100 mL 0.1 N sulfuric acid to 1 L using distilled water.

Actual process

Appropriate volume of filtered sample is taken out in 100 mL Nessler tube. 1 mL 1:1 HCl is added for acidification followed by the addition of 5 mL $BaCl_2$ solution. Stir well for 60 ± 2 s. After stirring period had ended, the solution is poured into absorption cell of photometer and absorbance is measured at 420 nm. Absorbance for the known sulfate concentrations had been derived in the same fashion using the standard sulfuric acid solution and graph is plotted for absorbance versus mg sulfate. The sample concentration is derived from the graph.[25,26]

Calculation

$$\text{Sulphate as } SO_4^{2-} \text{ mg/L} = \left[\text{mg } SO_4^{2-} \times 1000\right]/\text{mL sample} \qquad (2.38)$$

2.2.14 Phosphate

Phosphorous occurs in natural waters and in wastewater almost solely in the form of various types of phosphates. These forms are commonly classified into orthophosphates and total phosphates. These may occur in the soluble form, in particles of detritus, or in the bodies of aquatic organisms. Various forms of phosphates find their way into wastewater, effluents, and polluted water from a variety of sources. Larger quantities of the same compounds may be added when the water is used for laundering or other cleaning preparations, since these materials are major constituents of many commercial cleaning preparations. The presence of phosphate in large quantities in freshwaters indicates pollution through sewage and industrial wastes. It promotes growth of nuisance-causing microorganisms. Though phosphate possesses problems in surface waters, its presence is necessary for biological degradation of wastewaters.

Interferences and limitations

Arsenates in concentration of 100 mg/L react with the molybdate reagent to produce a blue color similar to that formed with phosphate. Silica in large concentration up to 10 mg/L interferes with the test. The presence of hexavalent chromium and NO^{2-} produces low results. NO^{2-} interference is avoided by adding sulfuric acid to the sample prior to the addition of ammonium molybdate, whereas chromium interference can be removed by adding ascorbic acid (1 mL) before the standard procedure.

Principle

In acidic conditions, orthophosphate reacts with ammonium molybdate to form molybdophosphoric acid. It is further reduced to molybdenum blue by adding a reducing agent such as stannous chloride or ascorbic acid. The blue color developed after the addition of ammonium molybdate is measured at 690 or 880 nm within 10-12 min after the development of color by using blank. The concentration is calculated from the standard graph. The intensity of the blue complex is measured, which is directly proportional to the concentration of phosphate present in the sample.

Apparatuses

a. Spectrophotometer (Boss and Lumps)
b. Nessler tubes having capacity of 100 mL

Reagents

i. *Stock phosphate solution*: dissolve 219.5 mg anhydrous KH_2PO_4 in distilled water and dilute to 1000 mL. 1 mL $= 50$ mg PO_4^{3-} P.

ii. *Phosphate working solution*: dilute 50 mL stock solution to 1000 mL with distilled water. 1 mL $= 2.50$ mg PO_4^{3-} P.

iii. *Ammonium molybdate solution*: dissolve 25 g ammonium molybdate in about 175 mL distilled water. Carefully add 280 mL concentrated H_2SO_4 to 400 mL distilled water. Cool and add molybdate solution and dilute to 1000 mL.

iv. *Strong acid reagent*: add 300 mL concentrated H_2SO_4 to 600 mL distilled water. Add 4 mL concentrated HNO_3, cool, and dilute to 1000 mL using distilled water.

v. *6 N sodium hydroxide*: take 24 g NaOH and dilute to 100 mL using distilled water.

vi. *Phenolphthalein*: dissolve 0.5 g in 500 mL 95% ethyl alcohol. Add 500 mL distilled water. Add dropwise 0.02 N NaOH until faint pink color appears (pH 8.3).

vii. *Stannous chloride reagent I*: dissolve 2.5 g fresh $SnCl_2 \cdot H_2O$ in 100 mL glycerol. Heat on water bath to ensure complete dissolution.

viii. *Dilute stannous chloride reagent II*: mix 8 mL stannous chloride reagent I with 50 mL glycerol and mix thoroughly.

ix. *Potassium antimonyl tartrate solution*: dissolve 2.7 g potassium antimonyl tartrate in 800 mL distilled water and dilute to 1000 mL.

x. *Ascorbic acid*: dissolve 1.76 g ascorbic acid in 100 mL distilled water. The solution is stable for a week at 4 °C.

xi. *Combined reagent*: mix 250 mL 5 N sulfuric acid, 75 mL ammonium molybdate solution, and 150 mL ascorbic acid solution. Add 25 mL potassium antimonyl tartrate solution and mix well. The solution must be prepared daily.

Actual process

1. *Orthophosphate*: take 50 mL sample into Nessler tubes using pipette. Add 4 mL ammonium molybdate followed by 0.5 mL stannous chloride or 8 mL combined reagent and dilute to 100 mL with distilled water and mix well. Allow to stand for 10 min. Prepare blank using distilled water in the same way. Measure the intensity of blue complex at 690 or 880 nm between 10 and 12 min after the development of the color. Calculate the PO_4^{3-} P (mg/L) from a calibration curve.

2. *Total acid hydrolyzable phosphate (total inorganic phosphate)*: take suitable volume of the sample in a conical flask. Add 1 drop of phenolphthalein. Add strong acid reagent until pink color disappears. Add 1 mL in excess. Boil for 5 min, cool, and filter if necessary. Transfer it to Nessler tube and neutralize to phenolphthalein with NaOH. Now, continue the process as per orthophosphate method. Measure the intensity of the complex at 690 or 880 nm and read the corresponding concentration from the calibration curve.

Polyphosphate
Calculation

$$\text{Total phosphate} = \text{Orthophosphate}$$
$$+ \text{Total inorganic phosphate} + \text{Polyphosphate} \quad (2.39)$$

2.2.15 Silica

Silica is the second most abundant element found on earth. Although silicon (Si) itself is a glassy insoluble solid, the various oxides (primarily "SiO_2") are somewhat soluble in water. It both exists as several minerals and is being produced synthetically. Notable examples include fused quartz, crystal, fumed silica, silica gel, and aerogels. Amorphous silica has a higher solubility of 100–140 ppm. When silica in the form of monosilicic acid ($Si(OH)_4$) is dissolved in water, it will remain in the monomeric state as long as its concentration remains less than about 2 mM. However, at higher concentrations, monosilicic acid dimerizes and polymerizes to form (larger) polysilicic acids, the larger of which are of a colloidal size. Colloidal silica is also sometimes referred to as silica. Higher amount of silica may have a problem in the evaporator.

Precautions
Collect samples in bottles of polyethylene, other plastics, or hard rubber. Glass is a less desirable choice, particularly in the case of waters with pH above 8.0 or with seawater, in which case a significant amount of silica in the glass can dissolve. Freezing to preserve samples can lower the soluble silica values by 20-40%. Do not acidify samples because silica precipitates in acidic condition.

Principle

Silica in water can be measured by gravimetric method and various color-imetry methods. We opted for molybdenum-blue method described below.

Ammonium molybdate at pH approximately 1.2 reacts with silica and any phosphate present to produce heteropoly acids. Oxalic acid is added to destroy the molybdophosphoric acid but not the molybdosilicate acid. Even if phosphate is known to be absent, the addition of oxalic acid is highly desirable and is a mandatory step in the method. The intensity of blue color is proportional to the concentration of "molybdate-reactive" silica. In at least one of its forms, silica does not react with molybdate even though it is capable of passing through filter paper and there is no noticeable turbidity. It is not known to what extent such "unreactive" silica occurs in waters.

Reagents

i. *1:1 HCl*: take 50 mL concentrate HCl and add 50 mL water.

ii. *10% ammonium molybdate solution*: take 10 g $(NH_4)_6Mo_7O_{24}\cdot4H_2O$ and dilute to 100 mL with distilled water.

iii. *Oxalic acid solution*: take 7.5 g oxalic acid and dilute to 100 mL with distilled water.

iv. *Stock silica solution*: take 4.73 g sodium metasilicate nonahydrate $(Na_2SiO_3\cdot9H_2O)$ and dilute to 1 L with distilled water.

v. *Standard silica solution*: take 10 mL of stock solution and dilute up to 1 L. 1 mL $= 0.01$ mg as SiO_2.

vi. *Reducing agent (amino naphthol sulfamic acid)*: take 500 mg 1-amino-2-naphthol-4-sulfonic acid and 1 g Na_2SO_3 in 50 mL distilled water. After warming, add it to 30 g $NaHSO_3$ in 150 mL distilled water. Filter it before use.

Actual process

Sample is taken out in a 100 mL Nessler tube. In rapid succession, 1.0 mL of 1:1 HCl is added for acidification and 2 mL 10% ammonium molybdate. This forms molybdic acid and molybdophosphate, if phosphate is present. After 2 min, add 2 mL 10% oxalic acid solution and mix thoroughly. Measure the time from the moment oxalic acid is added, wait at least 2 min but not more than 15 min, add 1 mL reducing agent, and mix thoroughly. This will reduce the molybdic acid (yellow) to heteropoly blue. After 5 min, the blue color is measured by visual comparison with the silica standards.[5]

2.2.16 Sodium

Sodium is a light alkali metal that will actually float on water in its pure state. However, it is very chemically active and is rarely found in its pure state. All natural waters contain sodium, as nearly all sodium compounds readily dissolve in water. Sodium is used primarily as sodium chloride (salt) and sodium sulfate (salt cake). Sodium is an essential biological chemical that takes part in processes maintaining osmotic pressure in living cells of plants and animals. An excess of sodium disturbs the critical balance, and in humans, excess salt (sodium chloride) intake has been linked to heart disease and high blood pressure. In plants, excess sodium leads to a perceived drought effect, and plants will show "burnt edge" effects and eventually die. Salt tolerance is a characteristic of many plants, but for agronomic purposes, including landscaping and wastewater reuse options, significant losses in plant quality and production arise from increased sodium in the plants' environment. Animals are tolerant to larger concentrations of sodium. It is common for sheep and cattle to eat soil, scraping the hard soil with their teeth, where salt extrusions occur along gully lines. Many animals also prefer salty groundwater to clean rainwater.

Principle

Trace amounts of sodium can be determined by flame emission photometry at wavelength 589 nm. The sample is sprayed into a gas flame and excitation is carried out under carefully controlled and reproducible conditions. The desired spectral line is isolated by the inbuilt filter. The intensity of light is measured by a phototube potentiometer. The intensity of light at 589 nm is approximately proportional to the concentration of element. The calibration curve may be linear but has a tendency to level off at higher concentrations.

Pretreatment

All samples were filtered through Whatman filter paper No. 1 to remove any suspended particles that otherwise may clog the capillary of the instrument.

Reagents

  i. *Stock sodium solution*: dissolve 2.542 g NaCl dried at 140 °C and dilute to 1 L with distilled water (1.0 mL = 1.0 mg Na).
 ii. *Intermediate sodium solution*: dilute 10 mL stock sodium solution to 100 mL with distilled water. This is 100 mg/L Na solution (1.0 mL = 0.1 mg Na).

iii. *Standard sodium solution*: dilute 10 mL of intermediate solution to 100 mL with distilled water. This is 10 mg/L Na solution (1.0 mL = 0.01 mg Na). Store all the solutions in plastic bottles. These solutions are used to prepare the calibration curve.

Actual process

Start the FPM Compressor 126. Start the Flame Photometer 128. Ignite the burner of the instrument. Keep the air pressure at 0.45 kg/cm², and adjust the gas feeder knob so as to have a blue sharp flame. In the software installed, first, we go to operating manual. The option for sodium measurement is selected. The instrument has two ranges inbuilt for sodium concentration measurement: one is 0–10 mg/L sodium and other is 0–100 mg/L sodium. The higher range was selected. Hence, sample dilution is required for samples containing sodium content higher than 100 mg/L. The dilution factor was also fed wherever applicable. Check the mg/L concentrations of the standards of 10 and 100 mg/L and calibrate the instrument by selecting option No. 5. After receiving "calibration over" message on the screen, the instrument is ready for measurement. Then, "Sampling" option, that is, option number 6, was selected. Filtered sample was poured in the beaker and capillary placed in it. Note the concentration (in mg/L) of sample on the screen. Aspirate distilled water after every measurement. No calibration curve is required here as the instrument has the advanced facility to directly convert the intensity into the concentration.

2.2.17 Heavy metals (Cu, Pb, Mn, and Cd)

AAS is a technique for measuring quantities of chemical elements, especially heavy metals, present in environmental samples by measuring the absorbed radiation by the chemical element of interest. This is done by reading the spectra produced when the sample is excited by radiation. The atoms absorb ultraviolet or visible light and make transitions to higher energy levels. The study of trace metals in wet and dry precipitation has increased in recent decades because trace metals have adverse environmental and human health effects. Heavy metals, such as chromium, lead, mercury, cadmium, and arsenic, are extremely toxic even in very small amounts. When any of these elements is present in the environment at high concentrations, living organisms are subjected to strong natural selection for tolerance. Environmental contamination by metals exerts physiological pressures that are clearly too severe for survival of most species by means of phenotypic plasticity or physiological acclimation, rather than genetic adaptation. Heavy metals are poisonous to living organisms including humans due to their biotoxic effects, which could be acute, chronic or subchronic, neurotoxic, carcinogenic, mutagenic, or teratogenic. Cadmium is toxic at extremely low levels; it is also associated with bone defects like osteomalacia, increased blood pressure, and myocardial dysfunctions. Severe exposure to cadmium may result in pulmonary edema and death. Smoking has also been reported to be a contributing factor to higher bioaccumulation of cadmium. Some scientists reported that lead is the most significant of the toxic heavy metals, and the inorganic forms are absorbed through ingestion of food and water as well as inhalation. Apart from the teratogenic effects of lead, its poisoning also causes inhibition of the synthesis of hemoglobin; dysfunctions in the kidneys, joints, and reproductive systems; acute and chronic damage to the central nervous system, etc. Workers with chronic headache and dizziness have higher levels of Cr and Pb in the scalp hair samples, such as in those working in a fireworks factory.

Principle

Atomic absorption methods measure the amount of energy in the form of photons of light that are absorbed by the sample. A detector measures the wavelengths of light transmitted by the sample and compares them to the wavelengths that originally passed through the sample. A signal processor then integrates the changes in the wavelength absorbed, which appear in the readout as peaks of energy absorption at discrete wavelengths. For detection of metals in AAS, two main components are required, i.e., first is the air (clean and dry air through a suitable filter to remove oil, water, and other foreign substances) and second is acetylene (standard commercial grade in cylinder).

Pretreatment

The samples were preserved by acidifying with concentrated HNO_3 to a pH of 2.0 or less and stored at approximately 4 °C to prevent change in volume due to evaporation. However, the measurement was done once the samples arrived at room temperature.

Samples were filtered through Whatman filter paper No. 1. Digestion of the samples was done, in which 5.0 g of thoroughly mixed homogeneous sample was taken out in a 100 mL glass beaker. 10 mL concentrated H_2SO_4 and 1 mL concentrated HNO_3 were added. They were then boiled on hot plate. One more mL of concentrated HNO_3 was added. Then on further boiling, 10 mL of 1:1 HCl is added. It is heated until almost 50% of the volume is lost. Wash down beaker walls with water and then filter. The filtrate was transferred to a 250 mL volumetric flask. Cool, dilute to mark, and mix thoroughly. A portion of this solution is taken out for required metal determination. The weight and volume data are fed in the instrument before the measurement.

Reagents
1. *1:1 HCl solution*: take 100 mL concentrated hydrochloric acid and add 100 mL distilled water.
2. *Standard metal solution (stock solution)*: the solutions are commercially available in the market (strength of copper 1 mL $= 100$ μg, lead 1 mL $= 100$ μg, manganese 1 mL $= 100$ μg, and cadmium 1 mL $= 100$ μg). All stock solutions by PerkinElmer are also used. Standard solutions of 1, 5, and 10 mg/L metal concentration were made from the stock solution.

Actual process
The following steps were followed for metal analysis on the PerkinElmer made instrument "AAnalyst 700." The very advanced instrument is attached to the PC (personal computer), and the whole program is run by the software "AAwinlab Analyst" version 4.1. On the instrument side, we have to start air and acetylene manually, and switch on the instrument. The instrument can hold eight lamps of different metals at a time. After having proper pressure of air and acetylene (75 and 30 kg/cm^2, respectively), ignite the burner. Autocalibration is done as the PC starts. Parameters involved in the detection of heavy metals in the program are mentioned in Table 2.1. The main program window gets open as the autocalibration is completed. Open "Method" from the file menu. This covers the lamp selection and lamp setup. The wavelength, slit, etc., details are taken automatically once the lamp is selected, which, however, can be changed if required. Open "Sample information file." The sample data like weight, volume, dilution, name, and date are inserted here. "Calibration graphics" involves setting atomization position for the required absorbance standard. Apply standard solution of 5 and 10 mg/L of a particular metal and a linear graph appears. If not, repeat the analysis, and recheck all the steps. Once the calibration curve is satisfactory, the instrument is now ready for sample measurement. Aspirate the sample blank. Aspirate distilled water after every measurement. Record the concentration in mg/L that appears directly on the screen.[5,6]

Table 2.1 Characteristics of Six Samples of Combined Wastewater of the Dyeing Mill Under Investigation

Parameters	July 2009	September 2009	November 2009	January 2010	March 2010	May 2010	July 2010	Mean
pH	8	7.84	8.21	7.65	8.21	9.24	8.25	8.2
COD (ppm)	1452	1478	1542	1620.2	1620.2	1642.8	1541.2	1556.6
BOD (ppm)	784	821	854	952.5	954.8	1002.4	1100.2	924.1
Color (Hazen)	241	233	251	321.2	341.2	350.2	301.2	291.3
TDS (ppm)	6554	6454	6888	6575	6758	7414	7274	6845.3
ES (µS/cm)	9874	9784	10,245	9998	1020	1124	9874	7417.0
Oil and grease (ppm)	12.5	14.5	8.25	14.5	15.45	12.5	14.5	13.2
Fluoride (ppm)	845	835	874	884	900	895	887	874.3
Sulfate (ppm)	821	811	898	900	941	915	854	877.1
Alkalinity (ppm)	658	644	724	828	500	671	670	670.7
Chloride (ppm)	445	411	458	487	514	500	512	475.3
Total hardness (ppm)	987	1000	1020	1102	1151	1202	1145	1086.7
Calcium hardness (ppm)	543	556	580	712	721	765	743	660
Magnesium hardness (ppm)	444	444	440	390	430	437	402	426.7
Phenol (ppm)	20.5	14.6	23.5	20.6	23.4	19.9	20.5	35.8
Phosphate (ppm)	20.4	25.4	33.4	35.3	25.5	27.1	33.3	50.1
Silica (ppm)	14.4	16.5	17.5	20.5	17.8	19.7	20.5	31.7
Sodium (ppm)	543.6	657.9	486.2	438.7	546.9	549.8	456.8	920.0
Copper (ppm)	2.3	1.4	3.2	1.9	2.5	2.1	2.3	2.24
Lead (ppm)	0.4	0.5	0.8	0.4	0.8	0.7	0.8	0.63
Manganese (ppm)	2.2	1.1	2.2	2.6	3.6	1.3	2.4	2.20
Chromium (ppm)	3.4	3.7	1.5	3.7	4.6	4.8	4.8	3.79

Month and year

2.3 CHARACTERIZATION

2.3.1 Results

The average values of various physicochemical parameters of wastewater sample collected bimonthly from July 2009 to July 2010 from a textile mill, GIDC Pandesara, Surat, before treatment processes, have been presented in Table 2.1. Also, Table 2.2 presents permissible limits of different parameters for wastewater of a textile mill before discharging them in nature, by different firms, i.e., Central Pollution Control Board (CPCB), US Environmental Protection Agency, and others, in which limits of some parameters are not available, but they are harmful to the environment if directly discharged into water bodies.

Table 2.2 Permissible Limits of Parameters for Textile Wastewater Given by Various Firms

	Permissible limits				
Parameters	CPCB	US EPA	World Bank Group	European board	FEPA[a]
pH	5.5-9.0	6.0-9.0	6.0-9.0	5.5-9.5	5.5-8.5
COD (ppm)	250	150	160	125	NA
BOD (ppm)	30	15	30	25	50
Color (Hazen)	100	NA	50	NA	NA
TDS (ppm)	NA	1500	NA	NA	2000
ES (µS/cm)	NA	NA	NA	NA	NA
Oil and grease (ppm)	10	10	10	NA	NA
Fluoride (ppm)	NA	NA	NA	25	NA
Sulfate (ppm)	1.0	0.5	1.0	500	NA
Alkalinity (ppm)	–	NA	NA	NA	NA
Chloride (ppm)	500	NA	NA	250	NA
Total hardness (ppm)	–	NA	NA	NA	NA
Calcium hardness (ppm)	–	NA	NA	NA	NA
Magnesium hardness (ppm)	–	NA	NA	NA	NA
Phenol (ppm)	1.0	0.5	0.5	NA	NA
Phosphate (ppm)	–	NA	2.0	10	5.0
Silica (ppm)	–	NA	NA	NA	NA
Sodium (ppm)	–	NA	NA	NA	NA
Copper (ppm)	–	NA	0.5	NA	1.0
Lead (ppm)	–	NA	NA	0.05	0.05
Manganese (ppm)	–	NA	NA	0.05	0.05
Chromium (ppm)	2.0	NA	0.5	0.05	0.05

[a]FEPA: Federal Environmental Protection Agency.

From Table 2.1, pH of the textile mill effluent ranges from 9.24 to 7.64, having an average value of 8.2, which is permissible. The color of the wastewater is light or dark yellow, having a color unit of 291.3 Hazen. As the permissible limit of color is 100 Hazen as per CPCB, so, wastewater must be treated to reduce color before discharging. The wastewater has COD and BOD values (average) of 1556.2 and 924.1 ppm, respectively, which shows that COD and BOD values are 5 and 10 times, respectively, higher than limits. Permissible limit of oil and grease is 10 ppm, which varies between 8.25 and 15.45 ppm in textile wastewater. The average value of sulfate is found to be 877.1 ppm, but its permissible limit is 100 ppm. Chloride in wastewater is found (475.3 ppm), which is under the permissible limit (500 ppm). The average value of phenol is 35.8 ppm, which is about 35 times higher than its limit. The average value of TDSs is found to be 6845.3 ppm. The average value of hardness and alkalinity values are 1086.7 and 670.7 ppm, respectively. The average values of fluoride and chloride are found to be 874.3 and 475.3 ppm, respectively. Also, the average value of EC is found to be 7417.0 μS/cm. The average value of the heavy metals (copper, lead, manganese, and chromium) are found to be 2.24, 0.63, 2.20, and 3.79 ppm, respectively, which are higher than the limits of the Federal Environmental Protection Agency (FEPA).

2.3.2 Discussion

It can be seen that the composition of the wastewater generated in an industry varies hourly as well as daily and the volume of effluent continuously increases from the start-up of the operation until its shutdown. It has been estimated that around 80% of the raw wastewater can be recycled through proper treatment by advance method.

Acids have marked effect on bacterial activity. pH influences the treatment methods and quality of water supply or wastewater. Therefore, the pH of the wastewater samples should be brought to the range of 7.6-9.2, before discharging. The pH value of the collected wastewater samples is suggested nearly acidic. The COD value of wastewater suggests the amount of oxygen required to oxidize the organic pollutants, whereas BOD values reflects the amount of oxygen required by bacterial population to stabilize the decomposable organic matter under aerobic conditions. These are not pollutants but indicators of pollution and can be considered an index of organic/inorganic chemicals present in the wastewater, and accordingly, suitable treatment methods can be finalized. The permissible limit for COD and

BOD is 100 and 30 ppm, respectively. As is evident from the analysis results, the COD value of wastewater of textile mill under study is nearly 15 times that of the limiting value. The BOD value is nearly 30 times that of tolerable value.

The presence of sulfates and chlorides depends widely on the methods and materials used in the production of textile and the quality of raw water. The TDSs are due to the chemicals used in the processing units and other operations, as well as other soluble organic and inorganic substances present in the samples. The combined wastewaters of textile industry show considerable amount of chloride indicating the presence of soluble salts containing chloride anion. The chloride compounds are well within the permissible limit, but TDSs, sulfate, and chloride are above the permissible limit.

Oil and grease content in the wastewater should be monitored carefully before discharging because it completely covers the surface of water and retards DO content of water. Extensive spreading of oil and grease affects the floating plantation and marine life severely and causes lethal toxicity on aquatic flora and fauna. The limiting value of oil and grease is up to 10 mg/L. Oil and grease content in the textile mill wastewater is well within the permissible limit. The presence of oil and grease in the effluent suggests the consumption of oil and grease by different units. All these results, revealed in the discussion below, fulfill the literature survey, mentioned in Tables 2.3–2.5.

Table 2.3 Characteristics of Effluents in M/s. Sivasakthi Textile Processors, Tirupur

S. no.	Parameters	Values
1	pH	9.76
2	Electrical conductivity (mS/cm)	6.80
3	Total COD (mg/L)	317
4	Total BOD5d (mg/L)	80
5	TDS (mg/L)	4280
6	TSS (mg/L)	47
7	Total hardness as $CaCO_3$ (mg/L)	320
8	Ca hardness as $CaCO_3$ (mg/L)	272
9	Sulfate (mg/L)	75
10	Chloride (mg/L)	1912
11	Sodium (mg/L)	1600
12	Potassium (mg/L)	38

Source: Ranganathan K, Karunagaran K, Sharma DC. Recycling of wastewaters of textile dyeing industries using advanced treatment technology and cost analysis—case studies. *Resour Conserv Recyl* 2007;**50**(3):306–18.

Table 2.4 Physicochemical Characteristics of Composite Wastewater of Small-Scale Textile Industry

S. no.	Parameters	8 h composite			24 h composite
		I[a]	II	III	
1	Temperature (°C)	20±2	25±3	26±4	25±4
2	Color (Pt-Co units)	830±110	790±130	840±170	820±180
3	Conductivity (mS/cm)	3.96±0.24	4.08±0.34	3.99±0.38	4.01±0.32
4	pH	9.5±0.6	9.4±0.7	9.6±0.5	9.5±0.6
5	Total alkalinity (mg/L)	1430±50	1420±40	1440±60	1430±60
6	Total dissolved solids (mg/L)	5070±268	5182±476	5098±362	5116±358
7	Total suspended solids (mg/L)	460±114	520±156	438±178	472±148
8	Ammoniacal nitrogen (mg/L)	5.0±0.2	6.1±0.3	7.3±0.4	6.1±0.2
9	Total Kjeldahl nitrogen (mg/L)	13.9±0.3	18.0±0.6	18.9±.04	16.7±0.5
10	Nitrate nitrogen (mg/L)	4.8±0.2	5.0±0.2	5.8±0.3	5.1±0.2
11	Chlorides (mg/L)	586±92	660±134	838±126	690±116
12	Sulfates (mg/L)	136±44	158±48	202±22	166±38
13	Phosphates (mg/L)	10.4±0.8	10.1±0.7	10.9±0.9	10.5±0.8
14	COD (mg/L)	714±98	770±112	798±104	760±102
15	BOD5d (mg/L)	198±42	225±45	220±65	215±50
16	Sodium (mg/L)	718±32	774±24	848±22	778±24

[a]Samples collected for 3 days with time interval of 1 h and composite for 8 h and 24 h.

Source: Pathe PP, Biswas AK, Rao NN, Kaul SN. Physicochemical treatment of wastewater from clusters of small scale cotton textile units. *Environ Technol* 2005;**26**:313-27.

Table 2.5 Physicochemical Characteristics of Effluents from the Textile Mills

S. no.	Parameters	Mill 1	Mill 2	Mill 3	Mill 4	Mill 5
1	Flow rate (m^3/day)	10,900	17,800	35,000	17,280	16,200
2	pH	10.21	11.23	11.04	11.53	10.47
3	Temperature (°C)	31.8	35.7	33.5	26.7	26.7
4	Color (Pt-Co)	2275	612	3537	4637	968
5	TDS (mg/L)	1130	2200	1480	848	250
6	TSS (mg/L)	245	35	471	1200	49
7	Sulfide (mg/L)	0.64	0.11	0.58	1.94	0.1
8	Free chlorine (mg/L)	0.01	0.01	1.14	1.06	0.76
9	COD (mg/L)	2120	1650	2430	2190	1067
10	BOD5d (mg/L)	227	382	645	242	163
11	Oil and grease (mg/L)	6.0	8.3	ND[a]	ND	ND
12	Dissolved oxygen (mg/L)	2.5	2.9	3.08	1.2	7.0
13	Nitrate (mg/L)	7.97	0.8	ND	ND	ND
14	Ammonia (mg/L)	1.82	2.01	1.28	0.05	2.72
15	Phosphate (mg/L)	3.42	0.09	2.63	0.74	0.36
16	Calcium (mg/L)	2.21	1.8	1.24	0.26	2.4
17	Magnesium (mg/L)	1.21	1.76	1.04	0.17	2.1

[a]ND, not detected.
Source: Yusuff RO, Sonibare JA. Characterization of textile industries' effluents in Kaduna, Nigeria and pollution implications. *Global Nest Int J* 2004;**6**(3):212-21.

REFERENCES

1. Bal AS. Wastewater management for textile industry—an overview. *Indian J Environ Health* 1999;**41**(4):264–90.
2. Lin SH, Chen L. Textile wastewater treatment by enhanced electrochemical method and ion exchange. *Environ Technol* 1997;**18**(7):739–46.
3. Rajagopalan S. Wastewater disposal problems in cotton textile industry, In: *Proceedings 16 technological conference, ATIRA, Ahmedabad*; 1975.
4. Lens PNL, Pol LH, Wilderer P, Asan T. *Water recycling and resource recovery in industry: analysis, technologies and implementation.* 1st ed. UK: IWA Publication; 2002.
5. APHA. *Standard methods for the examination of water and wastewater.* 19th ed. Washington, DC: American Public Health Association; 1999.
6. Maiti SK. *In: Water and wastewater analysis.* 1st ed. *Handbook of methods in environmental studies,* vol. 1. Jaipur, India: ABD Publishers; 2001.
7. Feldman I. Use and abuse of pH measurements. *Anal Chem* 1956;**28**:1859.
8. Rand MC, Greenberg AE, Taras MJ. *Standard methods for the examination of water and wastewater.* 14th ed. Washington, DC: APHA; 1976, p. 42-3.
9. Young JC, McDermott GN, Jenkins D. BOD determination. *J Water Sewage Works* 1981;**53**:1253.
10. TAPPI. *Standard and suggested methods.* New York: TAPPI; 1986.
11. Moore WA, Kroner RC, Ruchhoft CC. Dichromate reflux method for determination of oxygen consumed. *Anal Chem* 1949;**21**:953.
12. Pitwell LR, Standard COD. *Chem Br* 1983;**19**:90.

13. Kenny W, Resnick J, Personal communication with Young JC. Alterations in the BOD procedure for the 15th Edition of *Standard methods for the examination of water and wastewater*. J. WPCF, Davenport, Iowa; 1987.
14. Hammer MJ. BOD determination for industrial wastewater. *J Water Sewage Works* 1971;**118**(8):104–8.
15. Mackeow JJ, Brown LC, Gove GW. Comparative studies of dissolve oxygen analysis method. *J Water Sewage Works* 1967;**39**:1323.
16. Diehl H, Goetz CA, Hach CC. The versenate titration for total hardness. *J Am Water Works Assoc* 1950;**42**:40.
17. Goetz CA, Loomis TC, Diehl H. Total hardness in water: the stability of standard disodium dihydrogen ethylenediaminetetracetate solutions. *Anal Chem* 1950;**22**:798.
18. Barnard AJ, Broad WCJ, Flaschka H. The EDTA titration. *Chem Anal* 1956;**45**–86.
19. Goetz CA, Smith RC. Evaluation of various methods and reagents for total hardness and calcium hardness in water. *Iowa State J Sci* 1959;**34**:81.
20. Hazen A. On the determination of chloride in water. *J Am Chem* 1989;**11**:409.
21. Abbasi SA. Water Quality Sampling and Analysis. Discovery Publishing House, India, 1998.
22. Howard CS. Determination of total dissolved solids in water analysis. *Ind Eng Chem Anal Ed* 1933;**5**:4.
23. U.S. Geological Survey. *Methods for collection and analysis of water samples for dissolved minerals and gases. Techniques of water-resources investigation*. Washington, DC: U.S. Geological Survey; 1974 [Book 5, chapter A1].
24. Jenkins SR, Moore RC. A proposed modification to the classical method of calculating alkalinity in natural water. *J Am Water Works Assoc* 1977;**69**:56.
25. Kolthoff IM, Meeham EB, Sandell EB, Bruckenstein S. *Quantitative chemical analysis*. 4th ed. New York: Macmillan; 1969.
26. Rossum JR, Villarruz P. Suggested methods for turbidimetric determination of sulphate in water. *J Am Water Work Assoc* 1961;**53**:873.
27. Fould JM, Lunsford JV. New York. *J Water Sewage Works* 1968;**115**(3):112–5.
28. Schwarzenbach G, Flaschka H. *Complexometric titrations*. 2nd ed. Nueva York: Barnes and Nobel; 1969.

CHAPTER 3

Feasibility of Naturally Prepared Adsorbent

Contents

Abstract

This chapter starts with the basic introduction of adsorption, its types, and its mechanism. Investigated different sorbents for batch and column treatment are tabulated with respective references. Isotherms for batch (Freundlich and Langmuir) and columns (Thomas, Yoon-Nelson, bed depth service time, and Adams and Bohart) are

Characterization and Treatment of Textile Wastewater
http://dx.doi.org/10.1016/B978-0-12-802326-6.00003-4

discussed. Feasibility and comparison of naturally prepared adsorbents, that is, neem (*Azadirachta indica*) leaf powder, guava (*Psidium guajava*) leaf powder, and tamarind (*Tamarindus indica*) seed powder, and their activated forms using different acids are determined using various sophisticated analytic facilities like Fourier transform infrared, particle size distribution, scanning electron microcopy, and surface area, porosity, pore diameter, and pore volume analyses and also adsorptive batch treatment on dye solution. Treatment data using Freundlich and Langmuir isotherm are analyzed and compared. It is concluded that activated neem leaf powder, activated guava leaf powder using sulfuric acid, and normal tamarind seed powder are more suitable than their investigated analog adsorbents for the adsorption process to remove dyes and other contaminations.

Keywords: Adsorption, Naturally prepared adsorption, Analytic technique, Adsorption isotherm, Dye removal.

LIST OF FIGURES

LIST OF TABLES

3.1 INTRODUCTION

Adsorption is a surface phenomenon that is defined as the increase in concentration of a particular component at the surface of interface of the two phases. The term "adsorption" was proposed by Bios-Reymond but introduced into the world by Kayser.[1] The adsorption process has been widely used for the removal of solutes from solutions and harmful gases from the atmosphere. The adsorption process is efficient for the removal of organic matter from waste effluents. The adsorption takes place as a result of the removal of the solute from the solution. The concentration of the solute on the surface of a solid continues to increase until the solute in the solution remains in equilibrium with that at the surfaces.

The use of the term "sorption" instead of "adsorption" became common in the nineteenth century, for the surface activities. Sorption is defined as being the attraction of an aqueous species to the surface of a solid. Sorption is a rapid phenomenon of passive sequestration separation of sorbate from an aqueous/gaseous phase to a solid phase. Sorption occurs between two phases in transporting pollutants from one phase to another. It is considered to be a complex phenomenon and depends mostly on the surface chemistry or nature of the sorbent, sorbate, and system conditions in between the two phases. The amount of sorption that takes place on organic matter also follows various isotherms or kinetic rates. Sorption processes offer the most economical and effective treatment method for the removal of pollutants. The process is often carried out in a batch mode, by adding sorbent to a vessel containing contaminated water, stirring the mixture for a sufficient time and then letting the sorbent settle, and drawing off the cleansed water.

There are two types of sorption. If the sorption phenomenon is due to weak van der Waals forces, it is called physical sorption and is reversible in nature with low enthalpy values. On the other hand, in many systems, there may be a chemical bonding between sorbate and sorbent molecules. Such type of sorption is called chemisorption. As a result of chemical bonding, the sorption is irreversible in nature and has a high enthalpy of sorption.

Physical sorption (physisorption): Physisorption is relatively nonspecific and is due to the operation of weak forces between molecules. In this process, the sorbed molecule is not affixed to a particular site on the solid surface; it is free to move over the surface. The physical interactions among molecules, based on electrostatic forces, include dipole–dipole interactions, dispersion interactions, and hydrogen bonding. When there is a net separation of positive and negative charges within a molecule, it is said to have a dipole moment.

Molecules such as H_2O and N_2 have permanent dipoles because of the configuration of atoms and electrons within them. Hydrogen bonding is a special case of dipole–dipole interaction, and a hydrogen atom in a molecule has a partial positive charge. A positively charged hydrogen atom attracts an atom on another molecule that has a partial negative charge. When two neutral molecules that have no permanent dipoles approach each other, a weak polarization is induced because of interactions between the molecules, known as the dispersion interaction.[2]

In water treatment, the sorption of an organic sorbate from a polar solvent (water) to a nonpolar sorbent (carbonaceous material) has often been of interest. In general, attraction between sorbate and polar solvent is weaker for sorbates of a less polar nature; a nonpolar sorbate is less stabilized by dipole–dipole or hydrogen bonding to water. Nonpolar compounds are sorbed more strongly to nonpolar sorbents. This is known as hydrophobic bonding. Hydrophobic compounds sorb on to carbon more strongly. A longer hydrocarbon chain is more nonpolar, so the degree of this type of sorption increases with increasing molecular length.[2]

Chemical sorption (chemisorption): Chemisorption is also based on electrostatic forces, but much stronger forces have a major role in this process.[3] The attraction between sorbent and sorbate is a covalent or electrostatic chemical bond between atoms, with shorter bond length and higher bond energy in chemisorption.[2]

The enthalpy of chemisorption is very much greater than that of physisorption, and typical values are in the region of 200 kJ/mol, whereas this value for physisorption is about 20 kJ/mol. Except in special cases, chemisorption must be exothermic. A spontaneous process requires a negative free energy (ΔG) value. Because the translational freedom of the sorbate is reduced when it is sorbed, entropy (ΔS) is negative. Therefore, in order for ΔG to be negative, ΔH must be negative and the process exothermic. If the enthalpy values are less negative than -25 kJ/mol, the system is physisorption, and if the values are more negative than -40 kJ/mol, it is signified as chemisorption.[4]

Adsorption involves the interphase accumulation or concentration of substances at a surface or interface between two phases. The substance that is being adsorbed on the surface of another substance is called adsorbate. The substance present in bulk on the surface of which adsorption is taking place is called adsorbent. The interface may be liquid–liquid, liquid–solid, gas–liquid, or gas–solid. Of these types of adsorption, only liquid–solid adsorption is widely used in water and wastewater treatments. Adsorption is termed a surface phenomena, but actually, the whole mechanism does not take place at the surface; it also involves the area between the pore spaces of the adsorbent. The mechanism of sorption on the sorbent in the removal process involves the following four steps:

(1) Advective transport: movement of solute from bulk solutions to immobile film layer by means of advective flow or axial dispersion or diffusion.

(2) Film transfer: penetration and attachment of solute particles in immobile water film layer.

(3) Mass transfer: attachment of solute particle onto the surface of the adsorbent.

(4) Intraparticle diffusion: movement of solute into the pores of adsorbent and intraparticle diffusion.

Figure 3.1 illustrates the overall adsorption process, in which, out of four steps, the first step of mechanism is of negligible level in field, and the models to calculate the rate constants that control the next three mechanisms along are explained by Vasanth Kumar et al.[5]

As explained earlier, the sorption process is mainly controlled by mass transfer and adsorption equilibrium mechanisms. However, the real sorption process may get affected by the following factors:

(1) Competitive interaction between the different adsorbates [chemical oxygen demand (COD), color, heavy metals, and organic compounds] with adsorbent. The interaction between different adsorbates and adsorbents will reduce the sorption capacity of adsorbents for the particular pollutant to be removed. Competitive inhibition also interferes

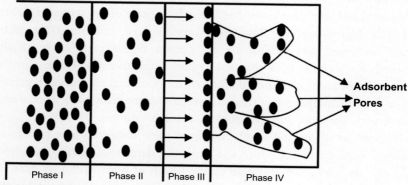

Figure 3.1 Different phases of adsorption phenomena.

with mass transfer kinetics of adsorbate and adsorbent. Also, there is no equilibrium model to better explain competitive inhibition.

(2) The presence of any sediments or suspected solids in the wastewater will cause fouling or choking in the adsorbent bed material. The presence of sediment material may also clog the pores of adsorbent material, thereby reducing the effective diffusivity and the mass transfer rate of the required solute to be removed.

(3) Estimating the significance of competitive inhibition and fouling factors for a particular field condition, it is possible to evaluate its impact on the adsorption capacity and kinetics of solute compound to be removed.

In the adsorption system, the contact between adsorbate and adsorbent mainly occurs with five different types: batch, continuous moving bed, continuous fixed bed (upflow or downflow), continuous fluidized bed, and pulsed bed. Sorption phenomena are dependent on experimental conditions like pH, temperature, sorbent dose, sorbent particle size, surface morphology of sorbent, sorbate concentration, types and structures of the sorbates, agitator speed, bed height, and contact time of adsorbent and adsorbate.[5] The surface morphology of sorbent is conducted by different analytic techniques such as Fourier transform infrared (FT-IR) spectroscopy, scanning electron microcopy (SEM), and porosity, pore diameter, pore volume, and surface area analyses. Many investigators have tried to study the surface chemistry of adsorbents using different analytic techniques.[6–25]

Various methods are available for the treatment of textile wastewater, namely, oxidation,[26–29] photocatalysis,[30] ozonation,[31,32] electrochemical method and ion exchange,[33–36] and membrane technology.[37] One of the most used processes for treatment of wastewater has been adsorption by activated carbon, an efficient solution. However, this treatment invloves high investment and operating costs, due to the high price of the activated carbon and to the high wastewater flow rate always involved, and these costs can be greatly increased when there are no carbon regeneration units locally. Research has recently been directed toward alternative adsorbents, namely, low-cost adsorbents, including the utilization of natural and waste materials. There is the incredible use of natural materials for the removal of contaminations from wastewater and dye from its aqueous solution.

The release of colored wastewater from these industries may present an ecotoxic hazard and introduce the potential danger of bioaccumulation. Dyes can also cause deterioration in the health of humans. Some dyes are found to be toxic, mutagenic, and carcinogenic. Dyes released by the industries can get into the water bodies and eventually contaminate the water supply system. Consumption of dye-polluted water can cause allergy reactions, dermatitis, skin irritation, cancer, and mutation in both babies and grown-ups. In addition, this problem can impact several vital activities such as fisheries, livestock, and agriculture since the polluted water is no longer suitable for their particular use.[38,39] Table 3.1 lists the various natural materials that are utilized for real and synthetic wastewaters (dye solution) (batch experiment).

3.1.1 Adsorption isotherm for batch treatment

The analysis and design of adsorption separation processes require the relevant adsorption equilibrium, which is the most important piece of information in understanding an adsorption process. It is also important for designing an adsorption system. The adsorption equilibrium indicates how the adsorbate molecules distribute between the liquid phase (solution) and the solid phase (adsorbent) when the adsorption process reaches an equilibrium state.[143] Adsorption isotherm is intended for determining the following:

(1)   The amount of adsorbate adsorbed per unit mass or unit area of the solid adsorbent, that is, the surface concentration of the adsorbate at a given temperature, since this is a measure of how much of the surface of the adsorbent has been covered and hence changed by the adsorption process.

(2)   The equilibrium concentration of adsorbate in the liquid phase required to produce a given surface concentration of adsorbate at a given temperature, since this measures the efficiency with which the adsorbate is adsorbed.

(3)   The concentration of adsorbate on the adsorbent at surface saturation at a given temperature, since this determines the effectiveness with which the adsorbate is adsorbed.

(4)   The orientation of the adsorbed adsorbate and any other parameters that may shed light on the mechanism by which the adsorbate is adsorbed, since a knowledge of the mechanism allows us to predict how an adsorbate with a given molecular structure will adsorb at the interface.

(5)   The effect of adsorption on other properties of the adsorbent.[144]

The most commonly used adsorption models are the Freundlich and Langmuir models to predict the sorption behavior.

Table 3.1 Natural Sorbents Used for Batch Adsorptive Treatment

Particulars	Name of adsorbent	Binomial name of origin	Reference no.
Textile wastewater	Sunflower stalks	*Helianthus annuus*	40
	Treated flute reed	*Phragmites karka*	41
	Bamboo-based activated carbon	*Phragmites karka*	42
	Coconut shell, coconut shell fibers, and rice husk	*Cocos nucifera* and *Oryza sativa*	43
Acid dyes	Acid-activated water hyacinth roots	*Eichhornia*	44
	Sunflower husk	*Helianthus annuus*	45
	Poly(methacrylic acid)–modified sugarcane bagasse	*Saccharum officinarum* L.	46
	Activated carbon from pistachio nut shell	*Pistacia vera*	47
	Lychee peel waste	*Litchi chinensis*	48
	Thermally activated spurge carbon	*Euphorbia macroclada*	49
	Activated carbons from sunflower seed hull	*Helianthus annuus*	50
	Activated carbon from poplar wood	*Populus*	51
	Carbon and activated carbon from waste coconut husks	*Cocos nucifera*	52
	Peanut hull	*Arachis hypogaea*	53
	Walnut, poplar, almond, and hazelnut	*Juglans regia, Populus, Prunus dulcis, Corylus avellana*	54
	Lemon peel	*Citrus limonum*	55
	Banana waste	*Musa* spp.	56
	Orange peel	*Citrus × sinensis*	57
Basic dyes	Activated carbon prepared from wood-apple shell	*Limonia acidissima*	58
	Activated carbon prepared from jackfruit peel waste	*Artocarpus heterophyllus*	59
	Gulmohar plant leaf powder	*Delonix regia*	60
	Tendu leaves	*Diospyros melanoxylon*	61
	Chiku	*Achras sapota*	62
	Acid-activated carbon prepared from *Pandanus*	*Pandanus utilis*	63
	Rice husk	*Oryza sativa*	64
	Pinecone of insignis pine	*Pinus radiata*	65

	Adsorbent	Scientific name	
66	Lignocellulosic material	*Saccharum officinarum* L.	
67	Dead leaves of neptune grass	*Posidonia oceanica* (L.)	
68	Pure and carbonized water hyacinth and water spinach	*Eichhornia crassipes* and *Ipomoea aquatica*	
69	Bagasse fly ash	*Saccharum officinarum* L.	
70	Chemically modified coriander	*Parthenium hysterophorus* L.	
71	Chinese flowering chestnut seed coat	*Xanthoceras sorbifolia*	
72	Oyster mushroom	*Pleurotus ostreatus*	
73	Canola hull	*Brassica campestris* L.	
74	Soy meal hull	*Glycine max*	
75	Coir pith and sugarcane fiber	*Cocos nucifera* and *Saccharum officinarum* L.	
76	Jute stick powder	*Corchorus capsularis*	
77	Coir pith carbon	*Cocos nucifera*	
78	Activated carbon and activated rice husks	*Oryza sativa*	
79	Giant duckweed	*Spirodela polyrhiza*	
80	Sargassum	*Sargassum binderi*	
81	Rubber seed coat	*Hevea brasiliensis*	
82	Pumpkin seed hull	*Cucurbita maxima*	
83	Spent tea leaves	*Camellia sinensis*	
84	Banana stalk waste	*Musa* spp.	
85	Coconut bunch waste	*Cocos nucifera*	
86	Pomelo peel	*Citrus grandis*	
87	Rice hull	*Oryza sativa*	
88	Dehydrated peanut hull	*Arachis hypogaea*	
89	Jute fiber carbon	*Corchorus capsularis*	
90	Rice bran and wheat bran	*Oryza sativa* and *Triticum aestivum*	
91	Palm-fruit bunch particles	*Cocos nucifera*	
92	Cocoa shell	*Cocos nucifera*	

Continued

Table 3.1 Natural Sorbents Used for Batch Adsorptive Treatment—Cont'd

Particulars	Name of adsorbent	Binomial name of origin	Reference no.
	Ground palm kernel coat	*Cocos nucifera*	93
	Almond shell	*Prunus dulcis*	94
	Palm shell-based activated carbon	*Cocos nucifera*	95
	Activated carbons prepared from rice husk	*Oryza sativa*	96
	Formaldehyde-treated parthenium biomass and phosphoric acid-treated parthenium carbon	*Parthenium incanum*	97
	Bagasse pith and maize cob	*Saccharum officinarum* L. and *Zea mays*	98
	Peanut hull	*Arachis hypogaea*	99
	Rice husk-based porous carbon	*Oryza sativa*	100
	Orthophosphoric acid-activated babul seed carbon	*Acacia nilotica*	101
Reactive dyes	Neptune grass fibrous	*Posidonia oceanica*	102
	Narrow-leaved cattails	*Typha angustifolia* L.	103
	Brazilian pine fruit wastes	*Araucaria angustifolia*	104
	Eucalyptus bark	*Eucalyptus globulus*	105
	Inactive mycelial biomass of *Panus fulvus*	*Panus fulvus*	106
	Positively charged functional group of the *Nymphaea rubra* biosorbent	*Nymphaea rubra*	107
	Brazilian pine fruit shell	*Araucaria angustifolia*	108
	Activated carbon from rice husk	*Oryza sativa*	109
	Almond shell	*Prunus dulcis*	94
	Activated carbon from coconut coir	*Cocos nucifera*	110
Direct dyes	Wood-apple fruit shell	*Feronia elephantum*	111
	Coconut husk	*Cocos nucifera*	112
	Orange peel carbon	*Citrus × sinensis*	113
	Surfactant-modified coconut coir pith	*Cocos nucifera*	114
	Activated carbon from palm kernel shell	*Cocos nucifera*	115

	Adsorbent	Scientific name	Ref.
	Activated carbon from orange peel	*Citrus × sinensis*	116
	Smooth loofah	*Luffa aegyptiaca*	117
	Palm ash	*Cocos nucifera*	118
	Almond shells	*Prunus dulcis*	119
	Activated carbon from orange peel	*Citrus × sinensis*	120
	Biomass of polypore mushroom	*Trametes versicolor*	121
Miscellaneous dyes	Rice husk ash	*Oryza sativa*	122
	Hydrogen peroxide–treated tendu waste	*Diospyros melanoxylon*	123
	Waste banana pith	*Musa* spp.	124
	Modified water nut carbon	*Eleocharis dulcis*	125
	Rice bran–based activated carbon	*Oryza sativa*	10
	Rice husk ash	*Oryza sativa*	126
	Peel of cucumber fruit	*Cucumis sativus*	127
	Natural and activated vine stem	*Vitis vinifera* L.	128
	Residue-based microwave-activated coffee press cake	*Coffea canephora*	129
	Chemically modified sunflower stalks	*Helianthus annuus*	130
	Activated carbon from bagasse pith	*Saccharum officinarum* L.	131
	Cellulosic waste orange peel	*Ananas comosus*	132
	Wheat husk	*Oryza sativa*	133
	Cellulosic waste orange peel	*Ananas comosus*	134
	Leaf biomass of apple of Sodom	*Calotropis procera*	135
	Beech wood sawdust	*Fagus grandifolia*	136
	Teakwood bark and rice husk	*Tectona grandis* and *Oryza sativa*	137
	Pericarp carbon from castor oil plant	*Ricinus communis*	138
	Rice husk ash	*Oryza sativa*	122
	Coconut husk	*Cocos nucifera*	139
	Mango leaves	*Mangifera indica*	141
	Peat	*Sphagnum girgensohnii*	142

Freundlich isotherm

The Freundlich isotherm was first used to describe gas phase adsorption and solute adsorption. Now it is widely used to understand adsorption phenomena. The empirical equation, based on multilayer sorption to a heterogeneous surface, of the Freundlich isotherm can be expressed as

$$q_e = K_F C_e^{1/n} \quad \text{or} \quad \log q_e = \log K_F + \left(\frac{1}{n}\right) \log C_e \qquad (3.1)$$

where q_e and C_e are the amount of adsorbed adsorbate per unit weight of adsorbent and unadsorbed adsorbate concentration in solution at equilibrium, respectively, and K_F and $1/n$ are the Freundlich constant characteristics of the system, which are determined from the $\log q_e$ versus $\log C_e$. One of the major disadvantages of the Freundlich equation is that it does not predict an adsorption maximum. The single K_F term in the Freundlich equation implies that the energy of adsorption on a homogenous surface is independent of surface coverage. Also, it can be derived theoretically by assuming the decrease in energy with the increasing coverage of the adsorbent surface. The researchers have often used the K_F and $1/n$ parameters to determine the mechanism of adsorption and have interpreted multiple slopes from Freundlich isotherms as evidence of different binding sites; such interpretation is speculative. It best describes sorption at lower amount of sorption on colloidal surface with the Freundlich isotherm.

Langmuir isotherm

Another widely used sorption model is the Langmuir equation. It was developed by Irving Langmuir[145] to describe the adsorption of a gas molecule on a planar surface. The assumptions for Langmuir are as follows: (i) Adsorption occurs on a planar surface that has a fixed number of sites that are identical and can hold only one molecule. Thus, monolayer coverage is permitted, which represents maximum adsorption. (ii) Adsorption is reversible. (iii) There is lateral movement of molecules on the surface. (iv) The adsorption energy is the same for all sites and independent of surface coverage (i.e., the surface is homogeneous), and there is no interaction between adsorbate molecules (i.e., the adsorbate behaves ideally). So, the Langmuir equation should only be used for purely qualitative and descriptive purposes. It was first applied to soil by Freid and Shapiro[146] and Olsen and Watanabe[147] to describe phosphate sorption on soil. Since that time, it has been heavily employed in many fields to describe sorption on colloidal surfaces.

The Langmuir adsorption equation can be expressed as

$$q_e = q_{max}K_L C_e/(1 + K_L C_e) \quad \text{or} \quad (1/q_e) = 1/q_{max} + \left[\frac{1}{(q_{max}K_L)}\right](1/C_e) \quad (3.2)$$

where C_e is the equilibrium concentration (mg/L); q_e is the quantity of dye adsorbed onto adsorbents (mg/g); q_{max} is q_e for a complete monolayer (mg/g), a constant related to sorption capacity; and K_L is the Langmuir constant related to the affinity of the binding sites and energy of adsorption (L/mg). The Langmuir parameters are obtained from the linear correlations between the values of $1/q_e$ and $1/C_e$.[148]

The batch treatment is useful in providing information about the effectiveness of dye–biosorbent system and sorption capacity parameter. However, the data obtained under batch conditions are generally not applicable to continuous systems (such as column operations) where contact time is not sufficient enough for the attainment of equilibrium. For the bulk removal of pollutants, continuous operation is often preferred over the batch operation, especially when dealing with large volumes of wastewater.[149] Some references are available for column studies for the removal of various contaminations of manmade wastewater (synthetic wastewater), which are tabulated in Table 3.2.

3.1.2 Column adsorption models

To design a column adsorption process, it is necessary to predict the breakthrough curve or concentration-like profile and adsorption capacity of the adsorbent for the selected adsorbate under the given set of operating conditions. It is also important for determining maximum sorption column capacity, which is a significant parameter for any sorption system. A number of mathematical models have been developed for the evaluation of efficiency and applicability of the column models for large-scale operations. The Thomas, Yoon-Nelson, bed depth service time (BDST), and Adams and Bohart models are most commonly used to analyze the behavior of adsorbent-adsorbate systems.

The Thomas model

The Thomas solution is one of the most general and widely used methods in column performance theory. The expression by the Thomas model[176] for an adsorption column is given as:

$$(C_t/C_0) = 1/[1 + \exp\{k_{TH}(q_0 x - C_0 v)/Q\}] \quad (3.3)$$

where C_t is the effluent dye concentration (mg/L), C_0 is the initial dye concentration (mg/L), x is the mass of the used adsorbent (g), v is the throughout

Table 3.2 Sorbents Used for Column Adsorptive Treatment

Adsorbent	Adsorbate	Reference no.
Zeolite 4A	Tetrahydrofuran	150
Activated carbon prepared from oil palm shell	Methylene blue dye	151
Activated carbon	Salicylic acid	152
Reaction mixtures	Palladium complexes	153
Activated carbon and activated slag	Malachite green	154
Clinoptilolite	Ammonium	155
Grape stalk wastes	Cadmium and lead	156
Titanium(IV) oxide nanoparticle	Ni(II) and Cr(VI)	157
Organoclays	Nitrobenzene	158
Surfactant-modified zeolite	Reactive yellow 176	159
Hydrolyzed wheat straw	Methylene blue and red basic 22	160
Brazilian coals	Carbon dioxide	161
Rice husk	Methylene blue	162
Granular activated carbon	Three components (COMPSORB-GAC)	163
Surface-tailored zeolite	Fluoride	164
Activated carbon	Mixed vapors of BTX	165
Duolite XAD–761 and Duolite A-568	α-Amylase	166
Green coconut shells	Cu(II), Pb(II), Cd(II), Zn(II), Ni(II)	167
Algal biomass	Cadmium	168
Untreated laterite	Arsenate	169
Poly(amidoamine) dendrimers	Lead	170
Phosphorus functionalized adsorbent	Neodymium(III)	171
Activated carbon	Basic dyes	172
Activated carbon by polyaniline	Arsenate	173
Activated carbon	Furfural and phenolic compounds	174

Adsorbent	Target	Ref.
Activated carbon	Reactive dyes	175
Aspergillus niger	Acid blue 29, basic blue 9, Congo red, and disperse red 1	176
Iron chromium oxide and lignite coal	Rhodamine B, Congo red and acid violet	177
Coconut husk	Methylene blue	112
Trametes versicolor or *Polyporus versicolor*	Phenol, 2-chlorophenol and 4-chlorophenol	178
Vermiculite	Cadmium	179
Perfil	4-Nitrophenol and 2,4-dinitrophenol	180
Base-treated cogon grass	Ni(II)	181
Ethylenediamine-modified rice hulls	Congo red	182
Granular activated carbon	Cr(VI)	183
Rubber seed shell	Methylene blue	184
Cation exchange resins	Ni(II)	185
Bark of *Pinus roxburghii*	Cr(VI)	186
Fly ash	Organic pollutants	187
Cellulose/chitin	Pb(II)	188
Anionic resin	Bilirubin	189
Fine-grained soil	Migratory nickel	190
Granular palygorskite	Phosphate	191
Gram husk and groundnut shell	Methylene blue, rhodamine B, congo red, eosine Y, and metanil yellow	192
Granular activated carbon (GAC F400)	Acid dyes	193
Granular activated carbon (Filtrasorb 400)	Acid dyes	194
Thermally activated *Euphorbia macroclada* carbon	Acid yellow 17 and acid orange 7	49
Sludge of the textile	Reactive red 2 and reactive red 141	195
Metal hydroxide sludge	Reactive red 141	196
Brilliant green	Acid-treated almond peel	197

volume of the dye solution (mL), and Q is the flow rate (mL/min). k_{TH} is the Thomas rate constant and q_0 is the maximum dye adsorption capacity of the adsorbent (mg/g), which is calculated from the plot of $\ln\left[(C_t/C_0) - 1\right]$ versus t.

The Yoon-Nelson model

The linear form of the Yoon–Nelson model is

$$\ln\left[\frac{C_t}{(C_0 - C_t)}\right] = k_{YN}t - \tau k_{YN} \qquad (3.4)$$

where k_{YN} is the Yoon-Nelson constant, τ is the time required for 50% adsorbate breakthrough, and t is the sampling time. A plot of $\ln\left[C_t/(C_0 - C_t)\right]$ versus t gives a straight-line curve with a slope of k_{YN} and intercept of $-\tau k_{YN}$. Based on τ, the adsorption capacity, q_{0YN}, is derived using

$$q_{0YN} = q_{(total)}/X = C_0 Q\tau/1000X \qquad (3.5)$$

So, adsorption capacity (q_{0YN}) related to the Yoon-Nelson model varies on inlet dye concentration (C_0), flow rate (Q), 50% breakthrough time derived from the Yoon-Nelson equation (τ), and weight of adsorbent (X).[198]

The BDST model

The BDST model relates the service time of a fixed bed with the bed height of adsorbent, with its quantity, because quantity is directly proportional to the bed height. The measurement of sorbent quantity is more precise than the determination of the respective volume, especially in the case of granules. Therefore, sorbent quantity is being preferably used, instead of the bed height. The linear form of the BDST model is

$$t = (N_0 Z/C_0 F) - [\{\ln(C_0/C_t) - 1\}/kC_0] \qquad (3.6)$$

where t is the service time (min), N_0 is the adsorption capacity (mg/g), F is the superficial liquid velocity (cm/min), Z is the height of the column (cm), and k is the rate constant of adsorption (L/min/mg) at time t. A plot of t versus bed depth, Z, should yield a straight line, where N_0 and k, the adsorption capacity and rate constant, respectively, can be evaluated.[199]

The Adams and Bohart model

Bohart and Adams established the fundamental equations that describe the relationship between C_t/C_0 and time in an open system. In spite of the fact that the original studies of Adams and Bohart were performed with the gas–charcoal adsorption system, their overall approach can be applied successfully in quantitative description of other systems. The model proposed assumes that the adsorption rate is proportional to both the residual capacity of the adsorbent and the concentration of the sorbing species. Assuming certain conditions, the linear form of this model[183] is

$$\ln\{(C_0/C_t) - 1\} = (K_{AB} N_0 Z/u) - K_{AB} C_0 t \qquad (3.7)$$

where K_{AB} is the kinetic constant, N_0 is the adsorption capacity coefficient, Z is the column height, and u is the linear velocity of fluid. The values of K_{AB} and N_0 are determined from the graph of $\ln[(C_0/C_t) - 1]$ versus t.

3.1.3 Present study

This chapter focuses on the preparation of naturally prepared adsorbents and their activation using different acids and analyzes the feasibility of naturally prepared adsorbents using various sophisticated analytic facilities like FT–IR spectrophotometry, particle size distribution, SEM, and surface area, pore volume, pore diameter, and porosity analyses. Also, it deals with comparison study of adsorbents and their activation by adsorptive batch treatment of different types of dyes and calculation of adsorption capacity.

3.2 MATERIALS AND METHODS

3.2.1 Preparation of adsorbent

Neem (scientific name: *Azadirachta indica*) belongs to the Meliaceae family and is native to the Indian subcontinent. Its seeds and leaves have been in use since ancient times to treat a number of human ailments and also as a household pesticide. Guava (*Psidium guajava*; family: Myrtaceae) is easily available in the Indian region.

The mature leaves of neem and guava used in the present investigation are collected from the available trees near Navyug Science College, Gujarat. The mature leaves of both plants are washed thrice with water to remove dust and water-soluble impurities and are dried until the leaves become crisp. The dried leaves are crushed and powdered and further washed with distilled water until the washings are free from color and turbidity. Then, this powder is dried in an oven at 60 ± 2 °C and placed in a desiccator for the adsorption studies, and thus, natural adsorbent prepared. Previously, neem leaf powder (NLP)[200–213] and guava materials (leaf and seed powder)[214–219] were utilized as adsorbents for the removal of various contaminations from their aqueous solution/wastewater by investigators. Also, the structure and chemical constitution of NLP was described by Ganguli.[220]

Tamarind (*Tamarindus indica*), from the family Fabaceae, has been used for the preparation of medicines for internal and external applications and as a condiment in many dishes. The major application of the seed commonly lies in the manufacturing of textile sizing powder. Tamarind fruit seeds, collected from nearby Navyug Science College, Gujarat, and a waste product of tamarind pulp are washed, dried, and pulverized. This powder is washed with distilled water until the washings are free from color and turbidity and, thereafter, dried in the oven for 2 h at 60 °C. References for the usage of tamarind and its derivatives as adsorbent are available.[221–234]

For the activation of adsorbent, each adsorbent is stirred with the excess amount of each acid for 30 min. These acids include 0.1 N sulfuric acid, 0.1 N hydrochloric acid, and 0.1 N nitric acid. Thereafter, it is washed with deionized water to remove untreated acid and dried in an oven at 60 ± 2 °C.

3.2.2 Surface chemistry of adsorbent

As discussed earlier, surface morphology of adsorbents plays a critical role in adsorption mechanism. Different sophisticated analytic techniques like FT-IR spectrophotometry, particle size distribution, surface area analysis, and SEM spectrometry technique are exploited to analyze the surface characterization of naturally prepared adsorbents after acid treatment. The instrument details used for the characterization of adsorbents are mentioned in Table 3.3.

Table 3.3 Instrument Details Used for Characterization of Natural Adsorbents and After Acid Activation

Sr. no.	Name of technique	Instrument maker (model no.)	Method
1	Fourier transform infrared (FT-IR) spectroscopy	Shimadzu, Japan (8400S)	KBr pellet method
2	Particle size analysis	Sympatic, Germany (Helos–BF)	–
3	Scanning electron microscopic analysis	Philips, the Netherlands (XL–30 ESEM)	ESEM method
4	Surface area analysis	Micromeritics (ASAP 2010)	BET method
5	Porosity, pore diameter, and pore volume analyses	Mercury porosimeter Thermo Quest (Pascal–140)	–

3.2.3 Adsorbate

Methylene blue dye (MBD) (CI No. 52015) is a heterocyclic aromatic chemical compound with the molecular formula of $C_{16}H_{18}N_3SCl$ and molecular weight of 319.85 g/mol. The dye is purchased from Sigma-Aldrich, India, and its structure is mentioned in Figure 3.2. This adsorbate is utilized for treatment with NLP and its activation forms.

The structure of synthetic textile dye Reactive Black 5 (RB5) (CI Reactive Black B-RB5) obtained from Merck, India, is illustrated in Figure 3.3. The molecular formula and weight are $C_{26}H_{25}N_5O_{19}S_6Na_4$ and 992 g/mol

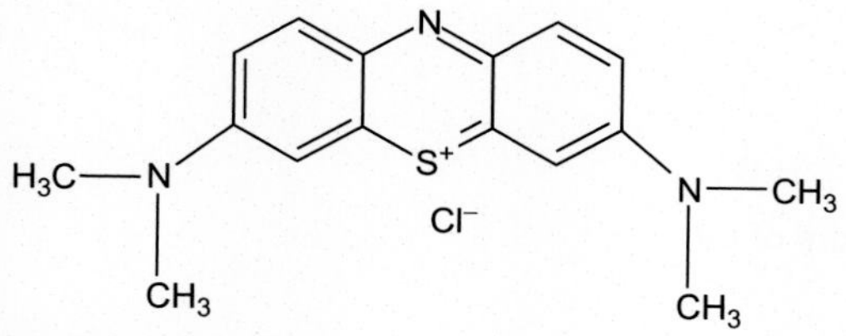

Figure 3.2 Structure of MBD.

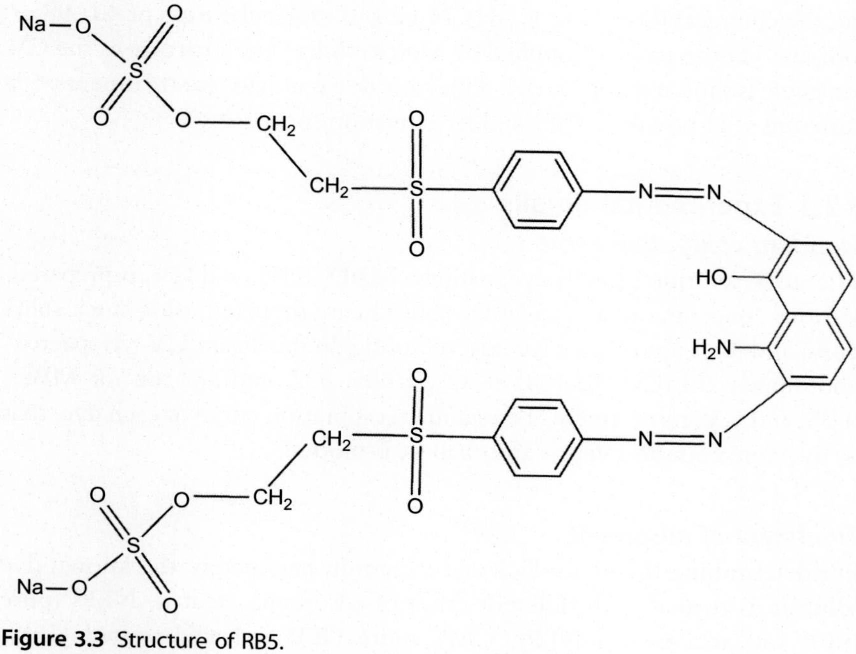

Figure 3.3 Structure of RB5.

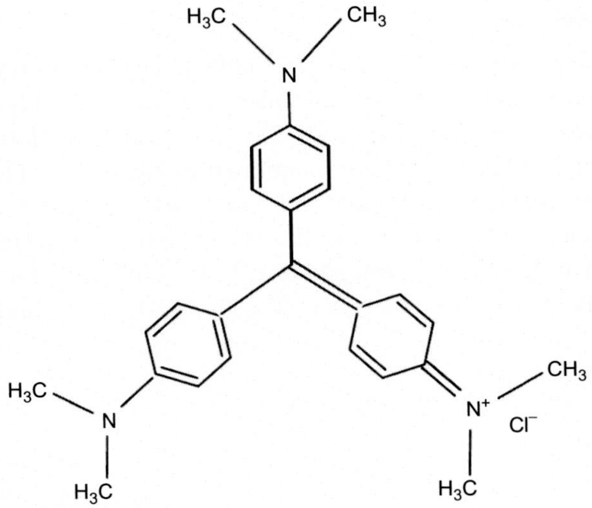

Figure 3.4 Structure of CV.

for RB5, respectively. RB5 is utilized for treatment with guava leaf powder (GLP) and its activation forms.

The dye, crystal violet (CV) (CI No. 42555; class: basic dye 3–CV) having the chemical formula of $C_{25}H_{30}N_3Cl$ and molecular mass of 407.98 g/mol, used in this study is supplied by Merck, India. The structure of the CV molecule is shown in Figure 3.4. This basic dye is utilized for treatment with tamarind seed powder (TSP) and its activation forms.

3.2.4 Experimental details
Standard calibration curve
The stock solution of each dye, namely, MBD, RB5, and CV, is prepared. Various concentrations of each dye solution are prepared using stock solutions, and their absorbance is measured using double beam UV–vis spectrophotometer (ELICO SL 164) at $\lambda_{max} = 665$, 592, and 584 nm for MBD, RB5, and CV, respectively. The standard calibration curve of each dye, that is, dye concentration versus absorbance, is plotted.

Treatment of adsorbent
For determining the Freundlich and Langmuir parameters, the known dye solution is treated with different doses of adsorbents, that is, NLPs (pure NLP and acid-treated NLP), GLPs (pure GLP and acid-treated GLP),

and TSPs (pure TSP and acid-treated TSP) at constant temperature, contact duration, and pH. Thereafter, the absorbance is measured using double beam UV-vis spectrophotometer (ELICO SL 164) at respective λ_{max}.

3.3 RESULTS AND DISCUSSION

3.3.1 Surface chemistry of adsorbent

As discussed earlier, surface morphology of adsorbents plays a critical role in adsorption mechanism. Different sophisticated analytic techniques like FFT-IR spectrophotometry, particle size distribution, SEM, and surface area, porosity, pore diameter, and pore volume analyses are exploited to analyze the surface characterization of naturally prepared adsorbents after acid treatment.

3.3.2 Fourier transform infrared (FT-IR) spectroscopy

The FT-IR spectroscopy displays a number of absorption peaks, which are used to investigate the presence of certain functional groups in a molecule. The FT-IR spectra of naturally prepared adsorbents and their acid-treated derivatives are recorded on a Shimadzu, Japan (8400S), spectrophotometer over the wavelength regions between 4000 and 400 cm^{-1}. Each of the spectra is a result of four scans. FT-IR spectra of samples are determined by using the potassium bromide disk technique. A small amount of finely ground biomass is mixed with about 100 times its weight of powdered potassium bromide. The mixture is thoroughly ground in a mortar and pestle and is then subjected to high pressure (18 psi) to form a small pellet about 1-2 mm thick and 1 cm in diameter. The resulting pellet is transparent and is used to test the surface functional groups by FT-IR spectroscopy.

The overlapped spectra of NLP, GLP, and TSP with their activation forms are taken. One of FT-IR spectra of GLP and its derivatives is mentioned in Figure 3.5. Table 3.4 represents FT-IR absorption bands and possible assignment of NLP, GLP, ad TSP and their activated forms. FT-IR studies revealed that the surface of naturally prepared adsorbents and their activation forms have various functional groups such as amino, hydroxyl, and carbonyl groups, which are generally used as adsorbents for the removal of various contaminations and dyes from water and wastewater stretching having large adsorption capacities of 80-90%.[235,10,13,45,69,236,11,73,47,23,237] Further, Gong et al.,[53] Liu et al.,[6] and Ahmad and Mondal[197] had reported that the hydroxyl group, —COOH/—COO$^-$, and ether group are important functional groups in the adsorption phenomena.

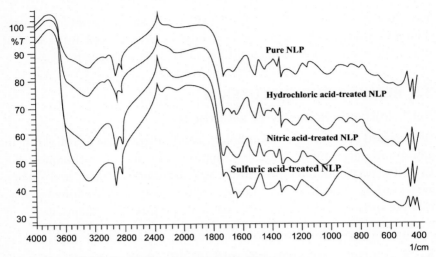

Figure 3.5 FT-IR spectra of GLP and its activated forms.

Table 3.4 FT-IR Absorption Bands of NLP, GLP, and TSP and Their Activated Forms

Sr. no.	Wavenumber (cm^{-1})	Possible assignment
1	3700–3200	Associated OH functional group of phenols, alcohols, and carboxylic acids (free and/or H–bonded OH group)
2	2900–2800	C—H stretching vibrations
3	1775–1700	Carbonyl compound, such as a ketone, an aldehyde, an ester, or a carboxylic acid
4	1550–1500	Aromatic ring vibrations
5	1350–1300	Methine C—H group, C—H stretching vibration
6	1250–1200	Aromatic primary amine, aromatic ethers, esters, and phenols
7	1100–1000	Secondary alcohol, C—O stretching, C—OH stretching vibration
8	450–400	Aromatic secondary amine, NH stretch

3.3.3 Scanning electron microscopy (SEM)

The SEM technique is used to determine the morphology of the adsorbent surface and is suitable for conductive surfaces. Areas ranging from approximately 1 cm to 5 μm in width can be imaged in a scanning mode using conventional SEM techniques (magnification ranging from 20 × to approximately 30,000 × and spatial resolution of 50–100 nm). The SEM is also capable of performing analyses of selected point locations on the sample; this approach is especially useful in qualitatively or semiquantitatively determining chemical compositions (using Energy Dispersive Spectroscopy (EDS)), crystalline structure, and crystal orientations (using Electron Back Scatter Diffraction (EBSD)). The SEM is routinely used to generate high-resolution images of shapes of objects (Scanning Electronic Imager (SEI)) and to show spatial variations in chemical compositions. An SEM is a type of electron microscope that images a sample by scanning it with a high-energy beam of electrons in a raster scan pattern. The electrons interact with the atoms that make up the sample-producing signals that contain information about the sample's surface topography. The accumulation of electric charge on the surfaces of nonmetallic specimens can be avoided by using environmental SEM in which the specimen is placed in an internal chamber at higher pressure, rather than the vacuum in the electron optical column. An Environmental Scanning Electronic Microscopy (ESEM) employs a scanned electron beam and electromagnetic lenses to focus and direct the beam on the specimen surface in an identical way as a conventional SEM. A very small focused electron spot (probe) is scanned in a raster form over a small specimen area. The beam electrons interact with the specimen surface layer and produce various signals that are collected with appropriate detectors. The output of these detectors modulates, via appropriate electronics, the screen of a monitor to form an image that corresponds to the small raster and information emanating from the specimen surface. In this study, the SEM test is conducted under a voltage of 20 kV and magnification range of 200–500.

The scanning electron microscopic images of pure NLP, sulfuric acid-activated NLP, nitric acid-activated NLP, and hydrochloric acid-activated NLP are conducted and analyzed, which reveals that the surface of sulfuric acid-activated NLP is more porous than that of other NLPs. Similarly, SEM micrographs of pure GLP, sulfuric acid-activated GLP, nitric acid-activated GLP, and hydrochloric acid-activated GLP demonstrated that GLP is found to be more porous after sulfuric acid treatment. But, scanning electron microscopic images of pure TSP, sulfuric acid-activated TSP, nitric acid-activated TSP, and hydrochloric acid-activated TSP are taken, which clearly indicate that the surface of TSP has no changes after acid (sulfuric acid, nitric acid, and hydrochloric acid) treatment.

3.3.4 Other surface analysis

Table 3.5 mentions the particle size, porosity, pore volume, pore diameter, and BET surface area analyses of naturally prepared adsorbents like NLP, GLP, and

Table 3.5 Surface Analysis of Adsorbents

Name of adsorbent	Surface area (m^2/g)	Particle size (mesh)	Porosity (%)	Pore volume (cm^3/g)	Ave. pore diameter (nm)
Pure NLP	457	124	24	0.052	8.5
Sulfuric acid–treated NLP	774	184	39	0.085	9.8
Nitric acid–treated NLP	540	160	30	0.074	8.9
Hydrochloric acid–treated NLP	510	132	28	0.065	8.9
Pure GLP	511	140	21	0.045	7.5
Sulfuric acid–treated GLP	741	178	35	0.074	8.9
Nitric acid–treated GLP	655	155	31	0.071	7.9
Hydrochloric acid–treated GLP	687	162	28	0.065	8.0
Pure TSP	574	122	20	0.038	6.5
Sulfuric acid–treated TSP	524	144	21	0.037	6.8
Nitric acid–treated TSP	478	133	24	0.040	6.5
Hydrochloric acid–treated TSP	501	128	19	0.034	6.4

TSP and their activated forms. Table 3.5 clearly represents that particle size, porosity, pore volume, pore diameter, and surface area of sulfuric acid–activated adsorbents are increased than those of other acids (nitric acid and hydrochloric acid) activated and normal adsorbent in case of NLP and GLP.

This shows that H_2SO_4 is effective in creating well-developed pores on the surface of NLP and GLP with large surface area and porous structure. But in the case of TSP, there is no change in the surface of TSP while it is activated using investigated acids.

3.3.5 Adsorption isotherm

The MBD solution having initial concentration of 200 mg/L is treated with NLP and various acid-activated NLPs (0.5–3.0 g/L) for adsorption MBD at temperature of 300 K and contact duration of 60 min. Experimental values obtained from Langmuir parameters, that is, adsorption capacity (K_L) and intensity (Q_0), and Freundlich parameters, that is, capacity (K_F) and intensity (n), are represented in Table 3.6.

The adsorption capacities related to Langmuir isotherm are found to be 401.6, 301.2, 352.9, and 352.6 L/g for sulfuric acid-treated NLP, nitric acid-treated NLP, hydrochloric acid-treated NLP, and pure NLP, respectively. So, we can say that sulfuric acid-treated NLP is found to be more efficient than other investigated NLPs. The parameters of hydrochloric acid-treated NLP and pure NLP are almost same, so there is no change in the surface of NLP when it is activated with sulfuric acid.

Table 3.6 Freundlich and Langmuir Parameters for the Adsorption of MBD Using Activated NLP and Pure NLP

Adsorbent	Freundlich parameters		Langmuir parameters	
	K_F (L/g)	n	K_L (L/g)	Q_0 (L/g)
Sulfuric acid-treated NLP	4.9936	1.43	0.0316	401.6
Nitric acid-treated NLP	4.9002	1.41	0.0345	301.2
Hydrochloric acid-treated NLP	4.8457	1.41	0.0381	352.9
Pure NLP	4.8397	1.46	0.0388	352.6

Temperature: 300 K, contact duration: 60 min, initial dye conc.: 200 mg/L.

The RB5 solution having an initial concentration of 60 mg/L is treated with GLP and various acid-activated GLPs (1.0–10.0 g/L) for adsorption RB5 at temperature of 300 K and contact duration of 4 h. Experimental values obtained for Langmuir parameters, that is, adsorption capacity (K_L) and intensity (Q_0), and Freundlich parameters, that is, capacity (K_F) and intensity (n), are depicted in Table 3.7. The maximum adsorption capacities (Q_{max}) of linear equation Langmuir isotherm are found to be 0.9282, 0.6584, 0.6125, and 0.5978 for sulfuric acid-treated GLP, nitric acid-treated GLP, hydrochloric acid-treated GLP, and pure GLP, respectively.

So, we can say that sulfuric acid-treated GLP is found to be more efficient than other investigated GLPs. The parameters of hydrochloric acid-treated GLP and pure GLP are almost same, so there is no change in the surface of NLP when it is activated with sulfuric acid.

The CV solution having the initial concentration of 6.0×10^{-6} M is treated with TSP and various acid-activated TSPs (1.0–10.0 g/L) for adsorption CV at temperature of 300 K and contact duration of 4 h. Table 3.8 demonstrates experimental values obtained for Langmuir parameters, that is, adsorption capacity (K_L) and intensity (Q_0), and Freundlich parameters, that is, capacity (K_F) and intensity (n).

Table 3.7 Freundlich and Langmuir Parameters for Adsorption of Reactive Green 12 Using Activated GLP and Pure GLP

Adsorbent	Freundlich parameters		Langmuir parameters	
	K_F (L/g)	n	K_L (L/g)	Q_0 (g/g)
Sulfuric acid-treated GLP	9.2726	1.033	0.02637	0.9282
Nitric acid-treated GLP	8.4512	1.784	0.03025	0.6584
Hydrochloric acid-treated GLP	7.4122	1.687	0.03125	0.6125
Pure GLP	7.3458	1.600	0.03145	0.5978

Temperature: 300 K, contact duration: 4 h, initial dye conc.: 60 mg/L.

Table 3.8 Freundlich and Langmuir Parameters for Adsorption of Crystal Violet Dye Using Activated TSP and Pure TSP

Adsorbent	Freundlich parameters		Langmuir parameters	
	K_F (mol/g)	n	K_L (mol/g)	Q_0 (mol/g)
Sulfuric acid-treated TSP	2.784×10^{-5}	1.448	0.047×10^{-5}	4.74×10^{-6}
Nitric acid-treated TSP	2.954×10^{-5}	1.844	0.087×10^{-5}	5.77×10^{-6}
Hydrochloric acid-treated TSP	2.124×10^{-5}	1.541	0.108×10^{-5}	6.07×10^{-6}
Pure TSP	2.814×10^{-5}	1.548	0.118×10^{-5}	6.87×10^{-6}

Temperature: 300 K, contact duration: 4 h, initial dye conc.: 6.0×10^{-6} M.

The maximum adsorption capacities (Q_{max}) of linear equation Langmuir isotherm are found to be 4.74×10^{-6}, 5.77×10^{-6}, 60.7×10^{-6}, and 6.87×10^{-6} mol/g for sulfuric acid-treated TSP, nitric acid-treated TSP, hydrochloric acid-treated TSP, and pure TSP, respectively. So, we can say that pure TSP is found to be more efficient than other investigated acid-treated TSPs. The parameters of hydrochloric acid-treated TSP and pure TSP are almost same, so there is no change in the surface of TSP when it is activated with sulfuric acid.

3.4 CONCLUSION

(1) The naturally prepared adsorbents, that is, NLP, GLP, and TSP, and also their activation forms using various acids are effectively utilized for the removal of dyes and various contaminations.

(2) Acids like sulfuric acid, nitric acid, and hydrochloric acid are effective in creating well-developed pores on the surface of naturally prepared adsorbents (NLP and GLP) obtained from sophisticated analyses of adsorbents.

(3) The activation of NLP and GLP using sulfuric acid is found more proficient than that of the regularly investigated NLPs and GLPs.

(4) Regular TSP is found to be more efficient than an activated TSP using sulfuric acid, nitric acid, and hydrochloric acid.

(5) The Freundlich and Langmuir parameters for natural and activated adsorbents are well fitted.

(6) Activated adsorbents (a–NLP and a–GLP) and TSP may be more convenient for the removal of COD, biochemical oxygen demand, and color from wastewater of dyeing mills.

REFERENCES

1. Dabrowski A. Adsorption—from theory to practice. *Adv Colloid Interface Sci* 2001;**93**:135–224.
2. Montgomery JM. *Water treatment principles and design.* USA: Consulting Engineers Inc.; 1985.
3. Sawyer NC, McCarty PL, Parkin GF. *Chemistry for environmental engineering.* Singapore: Mc-Graw Hill International Edition; 1994.
4. Atkins PV. *Physical chemistry.* 5th ed. Oxford: Oxford University Press; 1994.
5. Vasanth Kumar K, Subanandam K, Bhagavanulu DVS. Making GAC sorption economy. *Poll Res* 2004;**23**(3):439–44.
6. Liu Y, Wang W, Wang A. Removal of Congo red from aqueous solution by sorption on organified rectorite. *Clean Soil Air Water* 2010;**38**(7):670–7.
7. Jovic-Jovicic N, Milutinovic-Nikolic A, Grzetic I, Jovanovic D. Organobentonite as efficient textile dye sorbent. *Chem Eng Technol* 2008;**31**(4):567–74.
8. Saiah FBD, Su B, Bettahar N. Removal of Evans blue by using nickel-iron layered double hydroxide (LDH) nanoparticles: effect of hydrothermal treatment temperature on textural properties and dye adsorption. *Macromol Symp* 2008;**273**:125–34.
9. Yavuz E, Bayramoglu G, Arica MY, Senkal BF. Preparation of poly(acrylic acid) containing core-shell type resin for removal of basic dyes. *J Chem Technol Biotechnol* 2011;**86** (5):699–705.
10. Sankar M, Sekaran G, Sadulla S, Ramasami T. Removal of diazo and triphenylmethane dyes from aqueous solutions through an adsorption process. *J Chem Technol Biotechnol* 1999;**74**:337–44.
11. Liu Y, Wang W, Jin Y, Wang A. Adsorption behavior of methylene blue from aqueous solution by the hydrogel composites based on attapulgite. *Sep Sci Technol* 2011;**46** (5):858–68.
12. Monash P, Pugazhenthi G. Removal of crystal violet dye from aqueous solution using calcined and uncalcined mixed clay adsorbents. *Sep Sci Technol* 2010;**45**(1):94–104.
13. Yan Z, Li G, Muab L, Tao S. Pyridine-functionalized mesoporous silica as an efficient adsorbent for the removal of acid dyestuffs. *J Mater Chem* 2006;**16**:1717–25.
14. Priyantha N, Perera S. Removal of sulfate, phosphate and colored substances in wastewater effluents using feldspar. *Water Resour Manage* 2000;**14**:417–33.
15. Zhao D, Zhang Y, Wei Y, Gao H. Facile eco-friendly treatment of a dye wastewater mixture by in situ hybridization with growing calcium carbonate. *J Mater Chem* 2009;**19**:7239–44.
16. Bachri AM, Ali ZM. Removal of Congo red dye by adsorption onto phyrophyllite. *Int J Environ Stud* 2010;**67**(6):911–21.
17. Hernandez-Ramirez O, Holmes SM. Novel and modified materials for wastewater treatment applications. *J Mater Chem* 2008;**18**:2751–61.
18. Low LW, Teng TT, Ahmad A, Morad N, Wong YS. A novel pretreatment method of lignocellulosic material as adsorbent and kinetic study of dye waste adsorption. *Water Air Soil Pollut* 2011;**218**(1-4):293–306.
19. Ijagbemi CO, Chun JI, Han DH, Cho HY, Se OJ, Kim DS. Methylene blue adsorption from aqueous solution by activated carbon: effect of acidic and alkaline solution treatments. *J Environ Sci Health A* 2010;**45**(8):958–67.
20. Sathasivam K, Haris MRHM. Adsorption kinetics and capacity of fatty acid-modified banana trunk fibers for oil in water. *Water Air Soil Pollut* 2010;**213**:413–23.
21. Esparza P, Borges ME, Diaz L, Alvarez-Galvan MC, Fierro JLG. Equilibrium and kinetics of adsorption of methylene blue on Ti-modified volcanic ashes. *AIChE J* 2011;**57**(3):819–25.

22. Ahmad R, Kumar R. Adsorption of Amaranth dye onto alumina reinforced polystyrene. *Clean Soil Air Water* 2011;**39**(1):74–82.

23. Wang XS, Zhang W. Removal of basic dye crystal violet from aqueous solution by Cu(II)-loaded montmorillonite. *Sep Sci Technol* 2011;**46**(4):656–63.

24. Yuan X, Shi X, Zeng S, Wei Y. Activated carbons prepared from biogas residue: characterization and methylene blue adsorption capacity. *J Chem Technol Biotechnol* 2011;**86**:361–6.

25. Wang S, Yu D, Huang Y, Guo J. The adsorption of sulphonated azo-dyes methyl orange and xylenol orange by coagulation on hollow chitosan microsphere. *J Appl Polym Sci* 2011;**119**:2065–71.

26. Lei L, Hu X, Chen G, Porter JF, Yue PL. Wet air oxidation of desizing wastewater from the textile industry. *Ind Eng Chem Res* 2000;**39**:2896–901.

27. Amin H, Amer A, El Fecky A, Ibrahim I. Treatment of textile wastewater using H_2O_2/UV system. *Physicochem Probl Miner Process* 2008;**42**:17–28.

28. Radulescu C, Ionita I, Moater EI, Stihi C. Decolourization of textile wastewater containing green cationic dye by AOPs. *Ovidius Univ Ann Chem* 2009;**20**(1):66–71.

29. Balcioglu IA, Alaton IA, Otker M, Bahar R, Bakar N, Ikiz M. Application of advanced oxidation processes to different industrial wastewaters. *J Environ Sci Health A* 2003;**38**(8):1587–96.

30. Patil BN, Naik DB, Shrivastava VS. Treatment of textile dyeing and printing wastewater by semiconductor photocatalysis. *J Appl Sci Environ Sanit* 2010;**5**(3):309–16.

31. Perkowski J, Kos L, Ledakowicz S. Application of ozone in textile wastewater treatment. *Ozone Sci Eng* 1996;**18**(1):73–85.

32. Shang N, Chen Y, Yang Y, Chang C, Yu Y. Ozonation of dyes and textile wastewater in a rotating packed bed. *J Environ Sci Health A* 2006;**41**(10):2299–310.

33. Lin SH, Chen ML. Textile wastewater treatment by enhanced electrochemical method and ion exchange. *Environ Technol* 1997;**8**(7):739–46.

34. Muthukumar K, Sundaram PS, Anantharaman N, Basha CA. Treatment of textile dye wastewater by using an electrochemical bipolar disc stack reactor. *J Chem Technol Biotechnol* 2004;**79**:1135–41.

35. Mohan N, Balasubramanian N, Subramanian V. Electrochemical treatment of simulated textile effluent. *Chem Eng Technol* 2001;**24**:749–53.

36. Lidia S, Marta R. Electrochemical single-cell reactor for treatment of industrial wastewater composed of a spent textile bath. *Sep Sci Technol* 2007;**42**(7):1493–504.

37. Marcucci M, Ciabatti I, Matteucci A, Vernaglione G. Membrane technologies applied to textile wastewater treatment. *Ann N Y Acad Sci* 2003;**984**:53–64.

38. Chen KC, Wu JY, Huang CC, Liang YM, Hwang SCJ. Decolourization of azo dye using PVA-immobilized microorganisms. *J Biotechnol* 2003;**101**:241–52.

39. Garg VK, Gupta R, Yadav AB, Kumar R. Dye removal from aqueous solution by adsorption on treated sawdust. *Bioresour Technol* 2003;**89**:121–4.

40. Sun G, Xu X. Sunflower stalks as adsorbents for color removal from textile wastewater. *Ind Eng Chem Res* 1997;**36**:808–12.

41. Inthorn D, Tipprasertsin K, Thiravetyan P, Khan E. Color removal from textile wastewater by using treated flute reed in a fixed bed column. *J Environ Sci Health A* 2010;**45**(5):637–44.

42. Ahmad AA, Hameed BH. Reduction of COD and color of dyeing effluent from a cotton textile mill by adsorption onto bamboo-based activated carbon. *J Hazard Mater* 2009;**172**(2-3):1538–43.

43. Mohan D, Singh KP, Singh VK. Wastewater treatment using low cost activated carbons derived from agricultural byproducts—a case study. *J Hazard Mater* 2008;**152**(3):1045–53.

44. Rajamohan N. Equilibrium studies on sorption of an anionic dye onto acid activated water hyacinth roots. *Afr J Environ Sci Eng* 2009;**3**(11):399–404.

45. Soldatkina LM, Soldatkina EV, Menchuk VV. Adsorption of cationic dyes from aqueous solutions on sunflower husk. *J Water Chem Technol* 2009;**31**(4):238–43.

46. Xing Y, Wang G. Poly(methacrylic acid)-modified sugarcane bagasse for enhanced adsorption of cationic dye. *Environ Technol* 2009;**30**(6):611–9.

47. Vijayalakshmi P, Bala VSS, Thiruvengadaravi KV, Panneerselvam P, Palanichamy M, Sivanesan S. Removal of acid violet 17 from aqueous solutions by adsorption onto activated carbon prepared from pistachio nut shell. *Sep Sci Technol* 2011;**46**(1):155–63.

48. Bhatnagar A, Minocha AK. Assessment of the biosorption characteristics of lychee (*Litchi chinensis*) peel waste for the removal of Acid Blue 25 dye from water. *Environ Technol* 2010;**31**(1):97–105.

49. Gercel O, Gercel HF. Removal of acid dyes from aqueous solutions using chemically activated carbon. *Sep Sci Technol* 2009;**44**(9):2078–95.

50. Thinakaran N, Baskaralingam P, Ravi KVT, Panneerselvam P, Sivanesan S. Adsorptive removal of Acid Blue 15: equilibrium and kinetic study. *Clean* 2008;**36**(9):798–804.

51. Shakoohi R, Vatanpoor V, Zarrabi M, Varani A. Adsorption of Acid Red 18 (AR18) by activated carbon from poplar wood—a kinetic and equilibrium study. *E-J Chem* 2010;**7**(1):65–72.

52. Gupta VK, Jain R, Shrivastava M, Nayak A. Equilibrium and thermodynamic studies on the adsorption of the dye tartrazine onto waste "Coconut Husks" carbon and activated carbon. *J Chem Eng Data* 2010;**55**(11):478–84.

53. Gong R, Ding Y, Li M, Yang C, Liu H, Sun Y. Utilization of powdered peanut hull as biosorbent for removal of anionic dyes from aqueous solution. *Dyes Pigments* 2005;**64**(3):187–92.

54. Aydin AH, Bulut Y, Yavuz O. Acid dyes removal using low cost adsorbents. *Int J Environ Pollut* 2004;**21**(1):97–104.

55. Bhatnagar A, Kumar E, Minocha AK, Jeon B, Song H, Seo Y. Removal of anionic dyes from water using *Citrus limonum* (Lemon) peel: equilibrium studies and kinetic modeling. *Sep Sci Technol* 2009;**44**(2):316–34.

56. Kumar G, Ramalingam P, Kim M, Yoo C, Kumar M. Removal of acid dye (violet 54) and adsorption kinetics model of using *Musa* spp. waste: a low-cost natural sorbent material. *Korean J Chem Eng* 2010;**27**(5):1469–75.

57. Sivaraj R, Namasivayam C, Kadirvelu K. Orange peel as an adsorbent in the removal of Acid violet 17 (acid dye) from aqueous solutions. *Waste Manage* 2001;**21**(1):105–10.

58. Malarvizhi R, Sulochana N. Sorption isotherm and kinetic studies of methylene blue uptake onto activated carbon prepared from wood apple shell. *J Environ Prot Sci* 2008;**2**:40–6.

59. Prahas D, Kartika Y, Indraswati N, Ismadji S. The use of activated carbon prepared from jackfruit (*Artocarpus heterophyllus*) peel waste for methylene blue removal. *J Environ Prot Sci* 2008;**2**:1–10.

60. Ponnusami V, Aravindhan R, Karthikraj N, Ramadoss G, Srivastava SN. Adsorption of methylene blue onto gulmohar plant leaf powder: equilibrium kinetic and thermodynamic analysis. *J Environ Prot Sci* 2009;**3**:1–10.

61. Nagda GK, Ghole VS. Utilization of lignocellulosic waste from bidi industry for removal of dye from aqueous solution. *Int J Environ Res* 2008;**2**(4):385–90.

62. Rajasekhar KK, Shankarananth V, Srijanya C, Philo LMS, Reddy VL, Haribabu R. Adsorption studies of Congo red and methylene blue on the surface of *Achras sapota*. *J Pharm Res* 2009;**2**(9):1528–9.

63. Hema M, Arivoli S. Comparative study on the adsorption kinetics and thermodynamics of dyes onto acid activated low cost carbon. *Int J Phys Sci* 2007;**2**(1):10–7.

64. Mckay G, Porter JF, Prasad GR. The removal of dye colours from aqueous solution by adsorption on low cost materials. *Water Air Soil Pollut* 1999;**114**:423–38.
65. Sen TK, Afroze SA, Ang HM. Equilibrium, kinetics and mechanism of removal of methylene blue from aqueous solution by adsorption onto pine cone biomass of *Pinus radiate. Water Air Soil Pollut* 2010;**218**(1-4):499–515.
66. Low LWL, Teng TT, Ahmad A, Morad N, Wong YS. A novel pretreatment method of lignocellulosic material as adsorbent and kinetic study of dye waste adsorption. *Water Air Soil Pollut* 2010;**218**(1-4):293–306.
67. Cengiz S, Cavas LA. Promising evaluation method for dead leaves of *Posidonia oceanica* (L.) in the adsorption of methyl violet. *Mar Biotechnol* 2010;**12**:728–36.
68. Timi T, Michael Jr. H. Adsorption of methylene blue dye on pure and carbonized water weeds. *Bioremediat J* 2007;**11**(2):77–84.
69. Gupta VK, Mohan D, Sharma S, Sharma M. Removal of basic dyes (rhodamine B and methylene blue) from aqueous solutions using bagasse fly ash. *Sep Sci Technol* 2000;**35** (13):2097–113.
70. Lata H, Gupta RK, Garg VK. Removal of basic dye from aqueous solution using chemically modified *Parthenium hysterophorus. Linn. biomass Chem Eng Commun* 2008;**195** (10):1185–99.
71. Yao Z, Wang L, Qi J. Biosorption of methylene blue from aqueous solution using a bioenergy forest waste: *Xanthoceras sorbifolia* seed coat. *Clean* 2009;**37**(8):642–8.
72. Islek C, Sinag A, Akata I. Investigation of biosorption behavior of methylene blue on *Pleurotus ostreatus (Jacq) P Kumm. Clean* 2008;**36**(4):387–92.
73. Mahmoodi NM, Armai M, Bahrami H, Khorramfar S. The effect of pH on the removal of anionic dyes from colored textile wastewater using a biosorbent. *J Appl Polym Sci* 2011;**120**:2996–3003.
74. Mahmoodi NM, Arami M. Modeling and sensitivity analysis of dyes adsorption onto natural adsorbent from colored textile wastewater. *J Appl Polym Sci* 2008;**109**:4043–8.
75. Parab H, Sudersanan M, Shenoy N, Pathare T, Vaze B. Use of agro-industrial wastes for removal of basic dyes from aqueous solutions. *Clean* 2009;**37**(12):963–9.
76. Panda G, Das S, Guha A. Jute stick powder as a potential biomass for the removal of Congo red and rhodamine B from their aqueous solution. *J Hazard Mater* 2009;**164**:374–9.
77. Kavitha D, Navasivayam C. Experimental and kinetic studies on methylene blue adsorption by coir pith carbon. *Bioresour Technol* 2007;**97**:14–21.
78. Gupta VK, Jain R, Mathur M, Sikarwar S. Adsorption of Safranin-T from wastewater using waste materials-activated carbon and activated rice husks. *J Colloid Interface Sci* 2006;**303**:80–6.
79. Waranusantigul P, Pokethitiyook P, Kruatrachue M, Upatham ES. Kinetics of basic dye (methylene blue) biosorption by giant duckweed (*Spirodela polyrrhiza*). *Environ Pollut* 2003;**125**:385–92.
80. Tan P, Wong C, Ong S, Hi S. Equilibrium and kinetic studies for basic Yellow 11 removal by *Sargassum binderi. J Appl Sci* 2009;**9**:3005–12.
81. Hameed BH, Daud FBM. Adsorption studies of basic dye on activated carbon derived from agricultural waste: *Hevea brasiliensis* seed coat. *Chem Eng J* 2008;**139**(1):48–55.
82. Hameed BH, El-Khaiary MI. Removal of basic dye from aqueous medium using a novel agricultural waste material: pumpkin seed hull. *J Hazard Mater* 2008;**155**(3):601–9.
83. Hameed BH. Spent tea leaves: a new non-conventional and low-cost adsorbent for removal of basic dye from aqueous solutions. *J Hazard Mater* 2009;**161**(2-3):753–9.
84. Hameed BH, Mahmoud DK, Ahmad AL. Sorption equilibrium and kinetics of basic dye from aqueous solution using banana stalk waste. *J Hazard Mater* 2008;**158** (2-3):499–506.

85. Hameed BH, Mahmoud DK, Ahmad AL. Equilibrium modeling and kinetic studies on the adsorption of basic dye by a low cost adsorbent: coconut (*Cocos nucifera*) bunch waste. *J Hazard Mater* 2008;**158**(1):65–72.

86. Hameed BH, Mahmoud DK, Ahmad AL. Sorption of basic dye from aqueous solution by pomelo (*Citrus grandis*) peel in a batch system. *Colloids Surf A* 2008;**316**(1-3):78–84.

87. Ong ST, Lee CK, Zainal Z. Removal of basic and reactive dyes using ethylenediamine modified rice hull. *Bioresour Technol* 2007;**98**(15):2792–9.

88. Ozer D, Dursun G, Ozer A. Methylene blue adsorption from aqueous solution by dehydrated peanut hull. *J Hazard Mater* 2007;**144**(1-2):171–9.

89. Senthilkumar S, Varadarajan PR, Porkodi K, Subbhuram CV. Adsorption of methylene blue onto jute fiber carbon: kinetics and equilibrium studies. *J Colloid Interface Sci* 2005;**284**(1):78–82.

90. Wang XS, Zhou Y, Jiang Y, Sun C. The removal of basic dyes from aqueous solutions using agricultural by-products. *J Hazard Mater* 2008;**157**(2-3):374–85.

91. Nassar MM, Hamoda MF, Radwan GH. Adsorption equilibria of basic dyestuff onto palm-fruit bunch particles. *Water Sci Technol* 1995;**32**(11):27–32.

92. Theivarasu C, Mylsamy S, Sivakumar N. Cocoa shell as adsorbent for the removal of methylene blue from aqueous solution: kinetic and equilibrium study. *Universal J Environ Res Technol* 2011;**1**(1):70–8.

93. Oladoja NA, Aboluwoye CO, Oladimeji YB. Kinetics and isotherm studies on methylene blue adsorption onto ground palm kernel coat. *J Eng Environ Sci* 2008;**32**:303–12.

94. Duran C, Ozdes D, Gundogdu A, Senturk HB. Kinetics and isotherm analysis of basic dyes adsorption onto almond shell (*Prunus dulcis*) as a low cost adsorbent. *J Chem Eng Data* 2011;**56**(5):2136–47.

95. Mohammadi M, Hassani AJ, Mohamed AR, Najafpour GD. Removal of rhodamine B from aqueous solution using palm shell-based activated carbon: adsorption and kinetic studies. *J Chem Eng Data* 2010;**55**(12):5777–85.

96. Mohanty K, Naidu JT, Meikap BC, Biswas MN. Removal of crystal violet from wastewater by activated carbons prepared from rice husk. *Ind Eng Chem Res* 2006;**45** (14):5165–71.

97. Lata H, Mor S, Garg VK, Gupta RK. Removal of a dye from simulated wastewater by adsorption using treated parthenium biomass. *J Hazard Mater* 2008;**153** (1-2):213–20.

98. Nassar MM, El-Geundi MS. Comparative cost of colour removal from textile effluents using natural adsorbents. *J Chem Technol Biotechnol* 1991;**50**(2):257–64.

99. Gong R, Sun Y, Chen J, Liu H, Yang C. Effect of chemical modification on dye adsorption capacity of peanut hull. *Dyes Pigments* 2005;**67**(3):175–81.

100. Guo Y, Zhao J, Zhang H, Yang S, Qi J, Wang Z, et al. Use of rice husk-based porous carbon for adsorption of rhodamine B from aqueous solutions. *Dyes Pigments* 2005;**66** (2):123–8.

101. Sujatha M, Geetha A, Sivakumar P, Palanisamy PN. Orthophosphoric acid activated babul seed carbon as an adsorbent for the removal of methylene blue. *E-J Chem* 2008;**5**(4):742–53.

102. Ncibi MC, Mahjoub B, Seffen M. Adsorptive removal of textile reactive dye using *Posidonia oceanica* (L.) fibrous biomass. *Int J Environ Sci Technol* 2007;**4**(4):433–40.

103. Nilratnisakorn S, Thiravetyan P, Nakbanpote W. Synthetic reactive dye wastewater treatment by narrow-leaved cattails (*Typha angustifolia Linn*): effects of dye salinity and metals. *Sci Total Environ* 2007;**384**:67–76.

104. Royer B, Lima EC, Cardoso NF, Calvete T, Bruns RE. Statistical design of experiments for optimization of batch adsorption conditions for removal of reactive Red 194 textile dye from aqueous effluents. *Chem Eng Commun* 2010;**197**(5):775–90.

105. Morais LC, Goncalves EP, Vasconcelos LT, Beca CGG. Reactive dyes removal from wastewaters by adsorption on eucalyptus bark—adsorption equilibria. *Environ Technol* 2000;**21**(5):577–83.

106. Sathishkumar M, Binupriya AR, Vijayaraghavan K, Yun S. Two and three-parameter isothermal modeling for liquid-phase sorption of Procion blue H-B by inactive mycelial biomass of *Panus fulvus*. *J Chem Technol Biotechnol* 2007;**82**:389–98.

107. Renganathan S, Kalpana J, Kumar MD, Velan M. Equilibrium and kinetic studies on the removal of reactive red 2 dye from an aqueous solution using a positively charged functional group of the *Nymphaea rubra* biosorbent. *Clean* 2009;**37**(11):901–7.

108. Lima EC, Royer B, Vaghetii JCP, Simon NM, Cunha BMD, Pavan FA, et al. Application of Brazilian pine-fruit shell as a biosorbent to removal of reactive red 194 textile dye from aqueous solution: kinetics and equilibrium study. *J Hazard Mater* 2008;**155**:536–50.

109. Sharma YC. Adsorption characteristics of a low-cost activated carbon for the reclamation of colored effluents containing malachite green. *J Chem Eng Data* 2011;**56**(3):478–84.

110. Sharma YC, Uma, Upadhyay SN. Removal of a cationic dye from wastewaters by adsorption on activated carbon developed from coconut coir. *Energy Fuels* 2009;**23**(6):2983–8.

111. Chaudhari UE. Evaluation of adsorption efficiency of *Ferronia elefuntum* fruit shell for methylene blue from aqueous solution. *Asian J Chem* 2010;**22**(9):6722–8.

112. Low KS, Lee CK. The removal of cationic dyes using coconut husk as an adsorbent. *Pertanika* 1990;**13**:221–8.

113. Khaled A, El-Nemr A, El-Sikailya A, Abdelwahaba O. Treatment of artificial textile dye effluent containing Direct Yellow 12 by orange peel carbon. *Desalination* 2009;**238**(1–3):210–32.

114. Sureshkumar M, Namasivayam C. Adsorption behavior of Direct Red 12B and rhodamine B from water onto surfactant-modified coconut coir pith. *Colloids Surf A* 2008;**317**:277–83.

115. Nourouzi NM, Chuah TG, Choong TSY. Adsorption of reactive dyes by palm kernel shell activated carbon: application of film surface and film pore diffusion models. *E-J Chem* 2009;**6**(4):949–54.

116. El-Nemr A, Abdelwahab O, El-Sikaily A, Khaled A. Removal of Direct Blue-86 from aqueous solution by new activated carbon developed from orange peel. *J Hazard Mater* 2009;**161**(1):102–10.

117. El-Ashtoukhy EZ. *Loofa egyptiaca* as a novel adsorbent for removal of direct blue dye from aqueous solution. *J Environ Manage* 2009;**90**(8):2755–61.

118. Ahmad AA, Hameed BH, Aziz N. Adsorption of direct dye on palm ash: kinetic and equilibrium modeling. *J Hazard Mater* 2007;**141**(1):70–6.

119. Ardejani FD, Badii KN, Limaee Y, Shafaei SZ, Mirhabibi AR. Adsorption of Direct Red 80 dye from aqueous solution onto almond shells: effect of pH; initial concentration and shell type. *J Hazard Mater* 2008;**151**(2–3):730–7.

120. Khaled A, El-Nemr A, El-Sikaily A, Abdelwahab O. Removal of Direct N Blue-106 from artificial textile dye effluent using activated carbon from orange peel: adsorption isotherm and kinetic studies. *J Hazard Mater* 2009;**165**(1–3):100–10.

121. Bayramoglu G, Arıca MY. Biosorption of benzidine based textile dyes "Direct Blue 1 and Direct Red 128" using native and heat-treated biomass of *Trametes versicolor*. *J Hazard Mater* 2007;**143**(1–2):135–43.

122. Lakshmi UR, Srivastava VC, Mall ID, Lataye DH. Rice husk ash as an effective adsorbent: evaluation of adsorptive characteristics for Indigo Carmine dye. *J Environ Manage* 2009;**90**(2):710–20.

123. Nagda GK, Ghole VS. Biosorption of Congo red by hydrogen peroxide treated tendu waste. *Iranian J Environ Health Sci Eng* 2009;**6**(3):195–200.

124. Namasivayam C, Kanchana N. Removal of Congo red from aqueous solution by waste banana pith. *Pertanika J Sci Technol* 1993;**1**(1):33–42.

125. Ahmad R, Mondal PK. Application of modified water nut carbon as a sorbent in Congo red and malachite green dye contaminated wastewater remediation. *Sep Sci Technol* 2010;**45**(3):394–403.

126. Chowdhury AK, Sarkar AD, Bandyopadhyay A. Rice husk ash as a low cost adsorbent for the removal of methylene blue and Congo red in aqueous phases. *Clean* 2009;**37** (7):581–91.

127. Santhi T, Manonmani S. Malachite green removal from aqueous solution by the peel of *Cucumis sativa* fruit. *Clean Soil Air Water* 2011;**39**(2):162–70.

128. Ayan EM, Toptas A, Kibrislioglu G, Yalcinkaya EES, Yanik J. Biosorption of dyes by natural and activated vine stem interaction between biosorbent and dye. *Clean Soil Air Water* 2011;**39**(4):406–12.

129. Franca AS, Oliveira LS, Nunes AA. Malachite green adsorption by a residue-based microwave-activated adsorbent. *Clean Soil Air Water* 2010;**38**(9):843–9.

130. Shi W, Xu X, Sun G. Chemically modified sunflower stalks as adsorbents for color removal from textile wastewater. *J Appl Polym Sci* 1999;**71**:1841–50.

131. Gad HMH, El-Sayed AA. Activated carbon from agricultural by-products for the removal of rhodamine-B from aqueous solution. *J Hazard Mater* 2009;**168**:1070–81.

132. Namasivayam C, Muniaswamy N, Gayathri K, Rani M, Ranganathan K. Removal of dyes from aqueous solution by cellulosic waste orange peel. *Bioresour Technol* 1996;**57**:37–43.

133. Gupta VK, Jain R, Varshney S. Removal of Reactofix golden Yellow 3 RFN from aqueous solution using wheat husk—an agricultural waste. *J Hazard Mater* 2007;**142**:443–8.

134. Mohammed HJ, Kadhim BJ, Mohammed AS. Adsorption study of some sulphanilic azo dyes on charcoal. *E-J Chem* 2011;**8**(2):739–47.

135. Ali H, Muhammad SK. Biosorption of crystal violet from water on leaf biomass of *Calotropis procera*. *J Environ Sci Technol* 2008;**1**(3):143–50.

136. Dulman V, Cucu-Man SM. Sorption of some textile dyes by beech wood sawdust. *J Hazard Mater* 2009;**162**(2–3):1457–64.

137. McKay G, Ramprasad G, Mowli PP. Equilibrium studies for the adsorption of dyestuffs from aqueous solutions by low-cost materials. *Water Air Soil Pollut* 1986;**29**(3):273–83.

138. Madhavakrishnan S, Manickavasagam K, Vasanthakumar R, Rasappan K, Mohanraj R, Pattabhi S. Adsorption of crystal violet dye from aqueous solution using *Ricinus Communis* pericarp carbon as an adsorbent. *E-J Chem* 2009;**6**(4):1109–16.

139. Jain R, Shrivastava M. Adsorption studies of hazardous dye Tropaeoline 000 from an aqueous phase on to coconut-husk. *J Hazard Mater* 2008;**158**(2–3):549–56.

140. Hameed BH. Equilibrium and kinetics studies of methyl violet sorption by agricultural waste. *J Hazard Mater* 2008;**154**(1–3):204–12.

141. Mugugan T, Ganapathi A, Valliappan R. Removal of dyes from aqueous solution by adsorption on biomass of mango (*Mangifera indica*) leaves. *E-J Chem* 2010;**7**(3):669–76.

142. Sepulveda L, Fernandez K, Contreras E, Palma C. Adsorption of dyes using peat: equilibrium and kinetic studies. *Environ Technol* 2004;**25**(9):987–96.

143. Budyanto S, Soedjono S, Irawaty W, Indraswati N. Studies of adsorption equilibria and kinetics of amoxicillin from simulated wastewater using activated carbon and natural bentonite. *J Environ Prot Sci* 2008;**2**:72–80.

144. Rosen MJ. *Surfactants and interfacial phenomena*. 3rd ed. New Jersey: John Wiley & Sons Inc; 2004.

145. Langmuir I. The adsorption of gases on plane surfaces of glass, mica and platinum. *J Am Chem Soc* 1918;**40**:1361–403.
146. Freid M, Shapiro RE. Phosphate supply patterns of various soils. *Proc Soil Sci Soc Am* 1956;**20**:471–5.
147. Olsen SR, Watanabe FSA. Method to determine a phosphorus adsorption maximum of soils as measured by the Langmuir isotherm. *Proc Soil Sci Soc Am* 1957;**21**:144–9.
148. Sparks DL. *Environment soil chemistry*. 2nd ed. New York: Elsevier Science; 2003.
149. Banat F, Al-Asheh S, Al-Ahmad R, Bni-Khalid F. Batch-scale and packed bed sorption of methylene blue using treated olive pomace and charcoal. *Bioresour Technol* 2007;**98**:3017–25.
150. Liang H, Gao H, Kong Q, Chen Z. Adsorption of tetrahydrofuran + water solution mixtures by zeolite 4A in a fixed bed. *J Chem Eng Data* 2007;**52**:695–8.
151. Tan IAW, Ahmad AL, Hameed BH. Adsorption of basic dye using activated carbon prepared from oil palm shell: batch and fixed bed studies. *Desalination* 2008;**225**:13–28.
152. Polakovic M, Gorner T, Villieras F, Donato P, Bersillon JL. Kinetics of salicylic acid adsorption on activated carbon. *Langmuir* 2005;**21**:2988–96.
153. Girgis MJ, Kuczynski LE, Berberena SM, Boyd CA, Kubinski PL, Scherholz ML, et al. Removal of soluble palladium complexes from reaction mixtures by fixed-bed adsorption. *Org Process Res Dev* 2008;**12**:1209–17.
154. Gupta VK, Srivastava SK, Mohan D. Equilibrium uptake, sorption dynamics, process optimization, and column operations for the removal and recovery of malachite green from wastewater using activated carbon and activated slag. *Ind Eng Chem Res* 1997;**36**:2207–18.
155. Karadag D, Akkaya E, Demir A, Saral A, Turan M, Ozturk M. Ammonium removal from municipal landfill leachate by clinoptilolite bed columns: breakthrough modeling and error analysis. *Ind Eng Chem Res* 2008;**47**:9552–7.
156. Miralles N, Valderrama C, Casas I, Martınez M, Florido A. Cadmium and lead removal from aqueous solution by grape stalk wastes: modeling of a fixed-bed column. *J Chem Eng Data* 2010;**55**:3548–54.
157. Debnath S, Biswas K, Ghosh UC. Removal of Ni(II) and Cr(VI) with titanium(IV) oxide nanoparticle agglomerates in fixed-bed columns. *J Chem Eng Data* 2010;**49**:2031–9.
158. Patel HA, Bajaj HC, Jasra RV. Sorption of nitrobenzene from aqueous solution on organoclays in batch and fixed-bed systems. *J Chem Eng Data* 2009;**48**:1051–8.
159. Faki A, Turan M, Ozdemir O, Turan AZ. Analysis of fixed-bed column adsorption of reactive Yellow 176 onto surfactant-modified zeolite. *J Chem Eng Data* 2008;**47**:6999–7004.
160. Batzias F, Sidiras D, Schroeder E, Christina W. Simulation of dye adsorption on hydrolyzed wheat straw in batch and fixed-bed systems. *Chem Eng J* 2009;**148**:459–72.
161. Soares JL, Oberziner ALB, Jose HJ, Rodrigues A, Moreira RFPM. Carbon dioxide adsorption in Brazilian coals. *Energy Fuels* 2007;**21**:209–15.
162. Han R, Wang Y, Yu W, Zou W, Shi J, Liu H. Biosorption of methylene blue from aqueous solution by rice husk in a fixed-bed column. *J Hazard Mater* 2007;**141**:713–8.
163. Schideman LS, Marinas BJ, Snoeyink VL. Three-component competitive adsorption model for fixed-bed and moving-bed granular activated carbon adsorbers. Part II. model parameterization and verification. *Environ Sci Technol* 2006;**40**:6812–7.
164. Onyango MS, Leswifi TY, Ochieng A, Kuchar D, Otieno FO, Matsuda H. Breakthrough analysis for water defluoridation using surface-tailored zeolite in a fixed bed column. *Ind Eng Chem Res* 2009;**48**:931–7.
165. Yun J, Choi D, Kim S. Equilibria and dynamics for mixed vapors of BTX in an activated carbon bed. *AIChE J* 1999;**45**(4):751–60.

166. Bautista LF, Martınez M, Aracil J. Adsorption of α-amylase in a fixed bed: operating efficiency and kinetic modeling. *AIChE J* 2003;**49**(10):2631–41.
167. Sousa FW, Oliveira AG, Ribeiro JP, Rosa MF, Keukeleire D, Nascimento RF. Green coconut shells applied as adsorbent for removal of toxic metal ions using fixed-bed column technology. *J Environ Manage* 2010;**91**:1634–40.
168. Vilar VJP, Santos SCR, Martins RJE, Botelho CMS, Boaventura RAR. Cadmium uptake by algal biomass in batch and continuous (CSTR and packed bed column) adsorbers. *Biochem Eng J* 2008;**42**:276–89.
169. Maiti A, Gupta SD, Basu JK, De S. Batch and column study: adsorption of arsenate using untreated laterite as adsorbent. *Ind Eng Chem Res* 2008;**47**:1620–9.
170. Xu Y, Zhao D. Removal of lead from contaminated soils using poly(amidoamine) dendrimers. *Ind Eng Chem Res* 2006;**45**:1758–65.
171. Park H, Tavlarides LL. Adsorption of neodymium(III) from aqueous solutions using a phosphorus functionalized adsorbent. *Ind Eng Chem Res* 2010;**49**:12567–75.
172. El Qada EN, Allen SJ, Walker GM. Adsorption of basic dyes onto activated carbon using microcolumns. *Ind Eng Chem Res* 2006;**45**:6044–9.
173. Yang L, Wu S, Chen JP. Modification of activated carbon by polyaniline for enhanced adsorption of aqueous arsenate. *Ind Eng Chem Res* 2007;**46**:2133–40.
174. Sulaymon AH, Ahmed KW. Competitive adsorption of furfural and phenolic compounds onto activated carbon in fixed bed column. *Environ Sci Technol* 2008;**42**:392–7.
175. Al-Degsa YS, Khraisheh MAM, Allen SJ, Ahmad MN. Adsorption characteristics of reactive dyes in columns of activated carbon. *J Hazard Mater* 2009;**165**:944–9.
176. Fu Y, Viraraghavan T. Column studies for biosorption of dyes from aqueous solutions on immobilised *Aspergillus niger* fungal biomass. *Water SA* 2003;**294**:465–72.
177. Kannan N, Murugavel S. Column studies on the removal of dyes rhodamine-B, Congo red and acid violet by adsorption on various adsorbents. *Electron J Environ Agric Food Chem* 2007;**6**(3):1860–8.
178. Kumar NS, Boddu VM, Krishnaiah A. Biosorption of phenolic compounds by trametes versicolor polyporus fungus. *Adsorpt Sci Technol* 2009;**27**(1):31–46.
179. Mathialagan T, Viraraghavan T. Adsorption of cadmium from aqueous solutions by vermiculite. *Sep Sci Technol* 2003;**38**(1):57–76.
180. Yaneva Z, Marinkovski M, Markovska L, Meshko V, Koumanova B. Dynamic studies of nitrophenols sorption on Perfil in a fixed-bed column. *Maced J Chem Chem Eng* 2008;**27**(2):123–32.
181. Hanafiah NAKM, Zakaria H, Ngah WSW. Base treated Cogon grass (*Imperata cylindrica*) as an adsorbent for the removal of Ni(II): kinetic, isothermal and fixed-bed column studies. *Clean Soil Air Water* 2010;**38**(3):248–56.
182. Ong S, Tay EH, Ha S, Lee W, Keng P. Equilibrium and continuous flow studies on the sorption of Congo red using ethylenediamine modified rice hulls. *Int J Phys Sci* 2009;**4**(11):683–90.
183. Quintelas C, Fernandes B, Castro J, Figueiredo H, Tavares T. Biosorption of Cr(VI) by three different bacterial species supported on granular activated carbon—a comparative study. *J Hazard Mater* 2008;**153**:799–809.
184. Oladoja NA, Asia IO, Ademoroti CMA, Ogbewe OA. Removal of methylene blue from aqueous solution by rubber (*Hevea brasiliensis*) seed shell in a fixed-bed column. *Asia-Pac J Chem Eng* 2008;**3**:320–32.
185. Priya PG, Basha CA, Ramamurthi V. Removal of Ni(II) using cation exchange resins in packed bed column: prediction of breakthrough curves. *Clean Soil Air Water* 2011;**39**(1):88–94.
186. Ahmad R, Rao RAK, Masood MM. Removal and recovery of Cr(VI) from synthetic and industrial wastewater using bark of *Pinus roxburghii* as an adsorbent. *Water Qual Res J Can* 2005;**40**(4):462–8.

187. Sarkar M, Acharya PK, Bhattacharya B. Removal characteristics of some priority organic pollutants from water in a fixed bed fly ash column. *J Chem Technol Biotechnol* 2005;**80**:1349–55.
188. Zhou D, Zhang L, Zhou J, Guo S. Development of a fixed-bed column with cellulose/chitin beads to remove heavy-metal ions. *J Appl Polym Sci* 2004;**94**:684–91.
189. Piemonte V, Turchetti L, Annesini MC. Bilirubin removal from albumin-containing solutions: dynamic adsorption on anionic resin. *Asia-Pac J Chem Eng* 2010;**5**:708–13.
190. Ghosh SN, Mukherjee S, Kumar S, Chakraborty P, Fan M. Breakthrough adsorption study of migratory nickel in fine-grained soil. *Water Environ Res* 2007;**79**(9):1023–32.
191. Fangqun G, Jianmin Z, Huoyan W, Changwen D, Wenzhao Z, Xiaoqin C. Phosphate adsorption on granular palygorskite: batch and column studies. *Water Environ Res* 2011;**83**(2):147–53.
192. Chakrabarti S, Chaudhuri B, Dutta BK. Adsorption of model textile dyes from aqueous solutions using agricultural wastes as adsorbents equilibrium, kinetics and fixed bed column study. *Int J Environ Pollut* 2008;**34**(1-2):261–74.
193. Walker GM, Weatherley LR. Fixed bed adsorption of acid dyes onto activated carbon. *Environ Pollut* 1998;**99**(1):133–6.
194. Walker GM, Weatherley LR. Adsorption of acid dyes on to granular activated carbon in fixed beds. *Water Res* 1997;**31**(8):2093–101.
195. Vasques AR, de Souza SMGU, Valle JAB, de Souza U, Antonio A. Application of ecological adsorbent in the removal of reactive dyes from textile effluents. *J Chem Technol Biotechnol* 2009;**84**(8):1146–55.
196. Netpradit S, Thiravetyan P, Towprayoon S. Evaluation of metal hydroxide sludge for reactive dye adsorption in a fixed-bed column system. *Water Res* 2004;**38**(1):71–8.
197. Ahmad R, Mondal PK. Application of acid treated almond peel for removal and recovery of brilliant green from industrial wastewater by column operation. *Sep Sci Technol* 2009;**44**(7):1638–55.
198. Sivakumar P, Palanisamy PN. Adsorptive removal of Reactive and Direct dyes using non-conventional adsorbent—column studies. *Indian J Chem Technol* 2009;**68**:894–9.
199. Deliyanni EA, Peleka EN, Matis KA. Modelling the sorption of metal ions from aqueous solution by iron-based adsorbents. *J Hazard Mater* 2009;**172**(2-3):550–8.
200. Srivastava R, Rupainwar DC. A comparative evaluation for adsorption of dye on Neem bark and Mango bark powder. *Indian J Chem Technol* 2011;**18**:67–75.
201. Sharma A, Bhattacharyya KG. Adsorption of chromium (VI) on *Azadirachta indica* (neem) leaf powder. *Adsorption* 2004;**10**:327–38.
202. Athar M, Farooq U, Hussain B. *Azadirachata indicum* (neem): an effective biosorbent for the removal of lead (II) from aqueous solutions. *Bull Environ Contam Toxicol* 2007;**79**:288–92.
203. Khattri SD, Singh MK. Colour removal from synthetic dye wastewater using a bioadsorbent. *Water Air Soil Pollut* 2000;**120**:283–94.
204. Sharma SK, Mudhoo A, Jain G, Sharma J. Corrosion inhibition and adsorption properties of *Azadirachta indica* mature leaves extract as green inhibitor for mild steel in HNO_3. *Green Chem Lett Rev* 2010;**3**(1):7–15.
205. Rao PS, Reddy SKVN, Kalyani S, Krishnaiah A. Comparative sorption of copper and nickel from aqueous solutions by natural neem (*Azadirachta indica*) sawdust and acid treated sawdust. *Wood Sci Technol* 2007;**41**:427–42.
206. Babu BV, Gupta S. Adsorption of Cr(VI) using activated neem leaves: kinetic studies. *Adsorption* 2008;**14**:85–92.
207. Kumar SP. Removal of Congo red from aqueous solutions by neem saw dust carbon. *Colloid J* 2010;**72**(5):703–9.
208. Sharma YC, Uma. Optimization of parameters for adsorption of methylene blue on a low-cost activated carbon. *J Chem Eng Data* 2010;**55**:435–9.

209. Sundaram KMS, Curry J, Landmark M. Sorptive behavior of the neem–based biopesticide; azadirachtin; in sandy loam forest soil. *J Environ Sci Health B* 1995;**30** (6):827–39.

210. Mishra SP, Tiwari D, Prasad SK, Dubey RS, Mishra M. Biosorptive behavior of mango (*Mangifera indica*) and neem (*Azadirachta indica*) barks for [134]Cs from aqueous solutions: a radiotracer study. *J Radioanal Nucl Chem* 2007;**272**(2):371–9.

211. Sarma J, Sarma A, Bhattacharyya KG. Biosorption of commercial dyes on *Azadirachta indica* leaf powder: a case study with a basic dye rhodamine B. *Ind Eng Chem Res* 2008;**47**:5433–40.

212. Bhattarcharyya GK, Sarma A. Adsorption characteristics of the dye; brilliant green on neem leaf powder. *Dyes Pigments* 2003;**57**:211–22.

213. Immich APS, Mundim B, de Souza AAU, de Souza SMAGU. Color removal from textile effluent using *Azadirachta indica* leaf powder as an adsorbent. In: *Proceedings of European congress of chemical engineering (ECCE-6) Copenhagen, 16-20 September;* 2007.

214. Gaikwad RW, Kinldy SAM. Studies on auramine dye adsorption on psidium guava leaves. *Korean J Chem Eng* 2009;**26**(1):102–7.

215. Ponnusami V, Vikram S, Srivastava SN. Guava (*Psidium guajava*) leaf powder: novel adsorbent for removal of methylene blue from aqueous solutions. *J Hazard Mater* 2008;**152**(1):276–86.

216. Ponnusami V, Madhuram R, Krithika V, Srivastava SN. Effects of process variables on kinetics of methylene blue sorption onto untreated guava (*Psidium guajava*) leaf powder: statistical analysis. *Chem Eng J* 2008;**140**(1-3):609–13.

217. Abdelwahab O, El Sikaily A, Khaled A, El-Nemr A. Mass–transfer processes of chromium(VI) adsorption onto guava seeds. *Chem Ecol* 2007;**23**(1):73–85.

218. Elizalde-Gonzalez MP, Hernandez-Montoya V. Guava seed as an adsorbent and as a precursor of carbon for the adsorption of acid dyes. *Bioresour Technol* 2009;**100** (7):2111–7.

219. Lohani MB, Singh A, Rupainwar DC, Dhar DN. Studies on efficiency of guava (*Psidium guajava*) bark as bioadsorbent for removal of Hg (II) from aqueous solutions. *J Hazard Mater* 2008;**159**(2-3):626–9.

220. Ganguli S. Neem: a therapeutic for all seasons. *Curr Sci* 2002;**82**(11):1304.

221. Saha P, Chowdhury S, Gupta S, Kumar I, Kumar R. Assessment on the removal of malachite green using tamarind fruit shell as biosorbent. *Clean Soil Air Water* 2010;**38**(5-6):437–45.

222. Anirudhan TS, Radhakrishnan PG. Uptake and desorption of nickel(II) using polymerised tamarind fruit shell with acidic functional groups in aqueous environments. *Chem Ecol* 2010;**26**(2):93–109.

223. Anirudhan TS, Radhakrishnan PG. Adsorptive removal and recovery of U(VI), Cu(II), Zn(II) and Co(II) from water and industry effluents. *Bioremediat J* 2011;**15**(1):39–56.

224. Maiti A, Agarwal V, De S, Basu JK. Removal of As(V) using iron oxide impregnated carbon prepared from Tamarind hull. *J Environ Sci Health A* 2010;**45**:1207–16.

225. Anirudhan TS, Radhakrishnan PG, Suchithra PS. Adsorptive removal of mercury(II) ions from water and wastewater by polymerized tamarind fruit shell. *Sep Sci Technol* 2008;**43**:3522–44.

226. Saha P. Assessment on the removal of methylene blue dye using tamarind fruit shell as biosorbent. *Water Air Soil Pollut* 2010;**213**:287–99.

227. Sivasankar V, Ramachandramoorthy T, Chandramohan A. Fluoride removal from water using activated and MnO_2-coated Tamarind Fruit (*Tamarindus indica*) shell: batch and column studies. *J Hazard Mater* 2010;**177**:719–29.

228. Goud VV, Mohanty K, Rao MS, Jayakumar NS. Prediction of mass transfer coefficients in a packed bed using tamarind nut shell activated carbon to removal. *Phenol Chem Eng Technol* 2005;**28**(9):991–7.

229. Patil S, Deshmukh V, Renukdas S, Patel N. Kinetics of adsorption of crystal violet from aqueous solutions using different natural materials. *Int J Environ Sci* 2011;**1**(6):1116–34.
230. Abdullah MA, Devi Prasad AG. Biosorption of nickel (II) from aqueous solutions and electroplating wastewater using tamarind (*Tamarindus indica* L.) bark. *Aust J Basic Appl Sci* 2010;**4**(8):3591–601.
231. Verma A, Chakraborty S, Basu JK. Adsorption study of hexavalent chromium using tamarind hull-based adsorbents. *Sep Purif Technol* 2006;**50**(3):336–41.
232. Murugan M, Subramanian E. Studies on defluoridation of water by tamarind seed: an unconventional biosorbent. *J Water Health* 2006;**04**:453–61.
233. Ramadevi A, Srinivasan K. Agricultural solid waste for the removal of inorganics: adsorption of mercury(II) from aqueous solution by tamarind nut carbon. *Indian J Chem Technol* 2005;**12**:407–12.
234. Popuri R, Jammala S, Reddy KVNS, Abburi K. Biosorption of hexavalent chromium using tamarind (*Tamarindus indica*) fruit shell—a comparative study. *Electron J Biotechnol* 2007;**10**(3):358–67.
235. Amran MB, Zulfikar MA. Removal of Congo red dye by adsorption onto phyrophyllite. *Int J Environ Stud* 2010;**67**(6):911–21.
236. Yan Z, Tao S, Yin J, Li G. Mesoporous silicas functionalized with a high density of carboxylate groups as efficient absorbents for the removal of basic dyestuffs. *J Mater Chem* 2006;**16**:2347–53.
237. Wang XS, Chen JP. Removal of the azo dye Congo red from aqueous solutions by the marine alga *Porphyra yezoensis Ueda*. *Clean* 2009;**37**(10):793–8.

CHAPTER 4

Batch Adsorption Treatment of Textile Wastewater

Contents

Abstract

This chapter discusses the batch adsorption treatment of textile wastewater using naturally prepared adsorbents in order to remove COD, BOD, and color. The effects of various process parameters like adsorbent dose, contact duration, temperature, pH, and agitator speed are explored. All these adsorption data are analyzed using the Freundlich and Langmuir adsorption isotherm models. From the batch adsorption treatment, we concluded that activated neem leaf powder (a-NLP) is found to be the best among all adsorbents investigated for textile wastewater treatment. Maximum removal of COD (100%), BOD (98.9%), and color (100%) are attained at a temperature of 300 K with contact duration of 90 min and dosage of 25 g/L of a-NLP. The highest adsorption capacities (θ_0) related to the Langmuir isotherm are found to be 1.34, 0.82, and 0.31 mg/g for COD, BOD, and color, respectively, using a-NLP.

Keywords: Adsorption, Batch treatment, COD, BOD, Color, Batch isotherm.

LIST OF FIGURES

Figure 4.1 Freundlich plot for the removal of COD, BOD, and color by a–GLP.

Figure 4.2 Langmuir plot for the removal of COD, BOD, and color by a–GLP.

Figure 4.3 (a) Influence of temperature for the removal of COD, BOD, and color using a–NLP. (b) Influence of temperature for the removal of COD, BOD, and color using a–GLP. (c) Influence of temperature for the removal of COD, BOD, and color using TSP.

Characterization and Treatment of Textile Wastewater
http://dx.doi.org/10.1016/B978-0-12-802326-6.00004-6

LIST OF TABLES

4.1 INTRODUCTION

Textile industry effluents are a major source of water pollution because dyes, detergents, and other contaminants present in the wastewater undergo chemical and biological changes, consume dissolved oxygen, destroy aquatic life, and pose a threat to human health as many of these contaminants are highly toxic in nature. Adsorption has been found to be an efficient and economical process for the removal of pollutants such as dyes and color from wastewater. Several workers have studied the removal of dyes and color from wastewater using various adsorbents like low-cost natural materials, activated natural materials, and charcoal prepared from different natural materials. These low-cost natural materials are especially made from natural sources like plant root, leaf, seed, and peel. Considering single-step operation, batch adsorption treatment is one of the simplest techniques to investigate adsorption capacity and adsorption rate. The applications of batch adsorption treatment include the characterization of adsorbents and prediction of attenuation from pollutants and potential impact from wastewater. It can be used to measure adsorption equilibrium and kinetic mass transfer data. It is also useful in providing information about the effectiveness of dye–biosorbent system and sorption capacity parameter.

4.2 EXPERIMENTS

For the treatment of the wastewater sample, activated neem leaf powder (a–NLP), activated guava leaf powder (a–GLP), and tamarind seed powder

Table 4.1 Experimental Details for the Treatment of Wastewater Using Adsorbents

Effect of system	Adsorption dose (g/L)	Contact duration (min)	Temperature (K)	pH	Agitator speed (rpm)
Effect of adsorption dose	1, 5, 10, 15, 18, and 20	180	300	7	400
Effect of contact duration	5.0	60, 120, 150, 210, and 240	300	7	400
Effect of temperature	5.0	180	308, 313, 318, 323, and 328	7	400
Effect of pH	5.0	180	300	3, 5, 7, 9, and 11	400
Effect of agitator speed	5.0	180	300	7	200, 400, 500, 600, and 700

(TSP) having a particle size of 122–184 mesh are added to wastewater samples and the mixture is stirred. Each adsorbent is kept in contact until equilibrium state is attained. The experiments are carried out, as shown in Table 4.1. The required pH of the system is maintained by using 0.1 N HCl or 0.1 N NaOH during experiments. All chemicals used are of analytical reagent grade and purchased from Qualigens Fine Chemicals, India. The important physicochemical characteristics, that is, COD, BOD, and color, are determined before and after treatment using standard methods.

4.3 RESULTS AND DISCUSSION

Table 4.2 represents the effects of different doses of naturally prepared adsorbents, that is, a–NLP, a–GLP, and TSP, on various physicochemical characteristics of the wastewater, maintaining a temperature of 300 K, contact duration of 180 min, pH of 7, and agitator speed of 500 rpm.

It can be seen that there is a large reduction in the COD content when a–NLP, a–GLP, and TSP are used, from an initial value of 1625.8 ppm to zero for a–NLP; to 258.4 ppm for a–GLP; to 200.1 ppm at a dosage of 20 g/L. BOD had an initial value of 1002.4–11.2 ppm by 20 g/L of a–NLP; 1002.4–201.2 ppm by 20 g/L of a–GLP; and 1002.4–251.2 ppm by 20 g/L of TSP. At 20.0 g/L of a–NLP, decolorization of wastewater is achieved from

Table 4.2 Influence of Different Doses of Natural Adsorbents on Various Physicochemical Characteristics of Combined Wastewater

Treatment dose (g/L)	a-NLP			a-GLP			TSP		
	COD (ppm)	BOD (ppm)	Color (Hazen)	COD (ppm)	BOD (ppm)	Color (Hazen)	COD (ppm)	BOD (ppm)	Color (Hazen)
Untreated	1625.8	1002.4	350.2	1625.8	1002.4	350.2	1625.8	1002.4	350.2
1	1425.2	911.2	321.2	1521.2	951.2	339.5	1589.4	968.2	332.2
5	1085.3	649.7	264.85	1185.7	713.2	243.3	1250.2	733.95	264.5
10	512.1	302.8	131.2	658.5	410	150.2	611.4	578.4	201.4
15	75.3	75.7	27.6	356.3	258.3	90.8	300.4	320.4	142.9
18	21.4	11.2	3.5	258.4	201.2	75.2	200.1	251.2	100.2
20	NIL	11.2	NIL	258.4	201.2	75.2	200.1	251.2	100.2

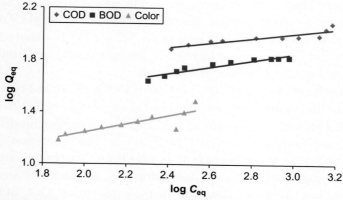

Figure 4.1 Freundlich plot for the removal of COD, BOD, and color by a-GLP.

an initial value of 350.2 Hazen. At 20.0 g/L of a-GLP, color can be removed up to 75.2 ppm from an initial value of 350.2 ppm, and at 20.0 g/L of TSP, up to 100.2 ppm from an initial value of 350.2 ppm. Figure 4.3a–c depicts the graphs of C_{eq} versus adsorbent dosage investigated for the removal of COD, BOD, and color using a–NLP, a–GLP, and TSP, respectively. The value of C_{eq} is continuously decreasing with increasing adsorbent dose up to 18.0 g/L.

The increase in adsorption with the increase in adsorbents may be attributed to the increased adsorbent surface and availability of more adsorption sites. Initially, the rate of increase in the percentage of components contributing to COD, BOD, and color removal is found to be rapid, which then slowed down as the dose increased. This phenomenon can be explained by the fact that at a lower adsorbent dose, the adsorbate is more easily accessible, resulting in higher removal per unit weight of adsorbent. The initial rise in adsorption with adsorbent or adsorbate concentration is probably due to a stronger driving force and larger surface area. The larger surface area of the adsorbent and the smaller size of adsorbate, increasing the active surface binding sites of the adsorbent, favor adsorption. Increasing concentration of certain functional groups of adsorbents like amino, hydroxyl, and carbonyl groups is also another reason for this phenomenon, which is discussed in Chapter 3.

Furthermore, the rate of adsorption is higher in the beginning as sites are available and the unimolecular layer increases. Adsorption and desorption occur together and the rates become equal at a stage called adsorption equilibrium when isotherms are applied. The subsequent slow rise in the curves is due to adsorption and intraparticle diffusion taking place simultaneously with the dominance of adsorption. With a rise in adsorbent dose, there is a less commensurate increase in adsorption resulting from lower adsorptive capacity utilization of adsorbents.

Removal per unit weight of adsorbent of COD, BOD, and color is calculated for the investigated naturally prepared adsorbents, and inverse values are needed for the conformation of the Langmuir adsorption model for the dose of adsorbents varied from 1 to 20 gm/L. The adsorption capacity decreases with the increasing amount of adsorbents. This may be attributed to overlapping or aggregation of adsorption sites, resulting in a decrease in total adsorbent surface area available to the particulars (COD, BOD, and color) and an increase in diffusion path length.

Freundlich isotherm (log C_{eq} vs. log q_e) and Langmuir isotherm (1/$C_{eq} \times 10^3$ vs. 1/$q_e \times 10^2$) for the removal of COD, BOD, and color from the wastewater using a-NLP, a-GLP, and TSP are plotted, and the linearity of the graph suggests the applicability of the adsorption model. Also, parameters of each model are calculated using intercept and slope. Freundlich and Langmuir plots for the removal of COD, BOD, and color from the wastewater using a–GLP are depicted in Figures 4.1 and 4.2, respectively.

Freundlich and Langmuir parameters for the removal of COD, BOD, and color from combined wastewater by adsorption onto naturally prepared adsorbents are depicted in Table 4.3, in which the values of n related to

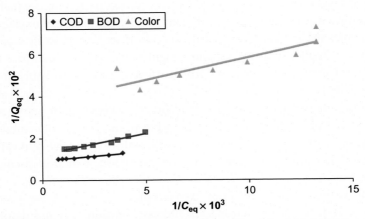

Figure 4.2 Langmuir plot for the removal of COD, BOD, and color by a-GLP.

Table 4.3 Freundlich and Langmuir Parameters for the Removal of COD, BOD, and Color from Combined Wastewater by Adsorption onto Naturally Prepared Adsorbents

Adsorbent	Particular	Freundlich parameter		Langmuir parameter	
		K_F (mg/g)	n	$K_L \times 10^3$ (L/mg)	q_{max} (mg/g)
a–NLP	COD	79.39	21.26	0.049	1.137
	BOD	28.69	10.42	0.123	1.142
	Color	23.07	8.92	0.170	1.444
a–GLP	COD	41.65	5.46	0.038	0.822
	BOD	12.14	3.91	0.141	0.782
	Color	11.70	3.11	0.174	0.476
TSP	COD	15.16	4.37	0.026	0.309
	BOD	3.98	4.41	0.066	0.288
	Color	2.03	2.55	0.157	0.281

adsorption intensities for COD are 21.26, 5.46, and 4.37 for a–NLP, a–GLP, and TSP, respectively.

The adsorption intensities for BOD are 10.42, 3.91, and 4.41 for a–NLP, a–GLP, and TSP, respectively. Further, adsorption intensities for color are 8.92, 3.11, and 2.55 for a–NLP, a–GLP, and TSP, respectively. The values of K related to adsorption capacities are found to be 79.39, 41.64, and 15.16 for COD, BOD, and color, respectively, when a–NLP is used. Further, the adsorption capacities are found to be 28.69, 12.14, and 3.98 for COD, BOD, and color, respectively, using a–GLP and 23.07, 11.70, and 2.03 for COD, BOD, and color, respectively, using TSP. The intercepts $K_L \times 10^3$ (L/mg) for COD, BOD, and color having value come to 0.049, 0.038, and 0.026, respectively, using a–NLP; 0.123, 0.141, and 0.066, respectively, using a–GLP; and 0.170, 0.175, and 0.157, respectively, using TSP. The adsorption capacity q_{max} is found to be 1.347, 0.822, and 0.309 for COD, BOD, and color, respectively, using a–NLP; 1.146, 0.782, and 0.288 for COD, BOD, and color, respectively, using a–GLP; and 1.444, 0.476, and 0.281 for COD, BOD, and color, respectively, using TSP.

Table 4.4 reveals the effects of different contact durations (60–240 min) on COD, BOD, and color removal using 5.0 g/L dosage of a–NLP, a–GLP, and TSP at constant temperature (300 K), pH 7, and agitator speed (400 rpm). The initial COD content is 1625.8 ppm and equilibrium concentration after 210 min of contact duration is found to be 598.8, 601.2, and 720.2 ppm using a–NLP, a–GLP, and TSP, respectively; initial BOD content, 1002.4 ppm and equilibrium concentration, 427.2, 504.2, and 600.2 ppm; and the initial

Table 4.4 Effect of Variation in Contact Duration of Naturally Prepared Adsorbents on Percent Removal of COD, BOD, and Color

Contact duration (min)	a-NLP			a-GLP			TSP		
	COD	BOD	Color	COD	BOD	Color	COD	BOD	Color
60	987.5	689.8	261.2	1112.2	715.4	311.2	1298.9	814.2	315.2
120	754.2	506.2	205.4	839.5	600.2	240.2	989.9	689.2	251.2
180	655.5	444.1	184.5	682.2	520.1	215.5	771.2	621.3	224.2
210	598.8	427.2	180.4	601.2	504.2	200.4	720.2	600.2	225.6
240	598.8	427.2	180.4	601.2	504.2	200.4	720.2	600.2	225.6

color content, 350.2 Hazen and equilibrium concentration, 180.4, 200.4, and 225.6 Hazen.

At the start, the rapid reduction in COD, BOD, and color is attributed to the presence of a large surface area and thus a greater number of active and vacant binding (adsorptive) sites on the outer surface of biomass, which results in the quick attachment of solute to the sorbent surface. After some time, the sorption begins to slow down due to the slow movement of solute molecules into the interior of the adsorbent bulk. But as the time proceeds, most of the active sites get bound with molecules contributing to COD, BOD, and color, and the number of free active sites decreases. It may decrease the number of successful collisions and hence decrease the rate of adsorption. In addition, there is a very high adsorption driving force in the beginning, which also results in a higher adsorption rate. However, at higher contact time, the rate of adsorption decreased and a saturation stage was attained due to the accumulation of the adsorption sites. This decline is due to the decrease in total adsorbent surface area and increased diffusion pathway.

Moreover, the amount of adsorption is found to increase with increasing contact time at all concentrations, and equilibrium is attained within about 210 min. It was further observed that the amount of metal ion uptake, q_t (mg/g), increases with increasing initial adsorbate concentration. The following are the three steps involved in the adsorption phenomenon in this kinetic experiment: a rapid adsorption at a shorter time (up to about 60 min), a transition phase, and an almost flat plateau section at the final stage (above 180 min). The first step is attributed to the fast utilization of the most readily available adsorbing sites on the adsorbent surface (bulk diffusion). The next step (up to about 180 min) exhibits additional removal, which is attributed to the diffusion of the adsorbate from the surface film to the macropores of the adsorbent (pore diffusion or intraparticle diffusion), stimulating further movement of the contaminations onto the adsorbent surface. The last is essentially an equilibrium step. For a solid–liquid adsorption process, the solute transfer is usually characterized by either external mass transfer (boundary layer diffusion) or intraparticle diffusion or both. The overall rate of sorption will be controlled by the slowest step, which would be either film or pore diffusion. However, the controlling step might also be distributed between intraparticle and external transport mechanisms.

Figure 4.3a–c depicts the effects of variation in temperature, that is, 298, 303, 308, 313, 318, 323, and 328 K, on adsorption by a-NLP, a-GLP, and TSP, respectively, at constant contact duration with respect to COD, BOD,

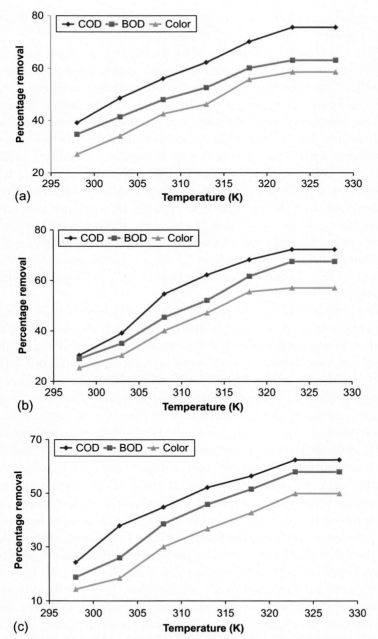

Figure 4.3 (a) Influence of temperature for the removal of COD, BOD, and color using a-NLP. (b) Influence of temperature for the removal of COD, BOD, and color using a-GLP. (c) Influence of temperature for the removal of COD, BOD, and color using TSP.

and color of wastewater. It can be seen that the increase in temperature leads to linear increases in percent removal in three cases. Also, a straight line after a temperature of 323 K indicates that equilibrium is attained at a temperature of 323 K. The values of percent removal of COD increase from 39.1% to 75.6% by a-NLP, from 30.3% to 72.3% by a-GLP, and from 24.3% to 62.5% by TSP; the values of percent removal of BOD, from 34.7% to 63.0% by a-NLP, from 29.0% to 67.9% by a-GLP, and from 18.7% to 58.1% by TSP; and the values of percent removal of color, from 27.1% to 58.5% by a-NLP, from 25.4% to 57.1% by a-GLP, and from 14.3% to 50.0% by TSP. All these removals are found at a temperature of 298–328 K. The adsorption capacity increases with the increasing temperature, indicating that adsorption is an endothermic process.

Temperature is also an important parameter to the adsorption process. The bigger adsorptive capacity of dyes was observed in the higher temperature range. The overall adsorption process consists of several steps: bulk film diffusion, intraparticle/pore diffusion, adsorption, and desorption in temperature affect the diffusion section. Increasing the temperature increases the rate of diffusion of the adsorbate molecules across the external boundary layer and in the internal pores of the adsorbent particle, owing to the decrease in the viscosity of the solution. Thus, a change in temperature alters the equilibrium capacity of the adsorbents for a particular adsorbate. Also, the solubility and chemical potential of adsorbate are related to temperature; higher temperatures facilitate the adsorption of pollutants on naturally prepared adsorbent because the mobility of the substance containing pollutants increases with rising temperature and the substance interacts more effectively with the functional groups (to enhance the rate of protonation and deprotonation) on naturally prepared adsorbent. At equilibrium (at much higher temperature), the decrease in adsorption may be due to the weakening of binding forces between adsorbate and adsorbent.

Increasing the temperature produces a swelling effect within the internal structure of naturally prepared adsorbent. Due to the swelling effect, an enlargement of the pores occurs and there is greater availability of molecules with enough energy to interact with the active sites on the surface of adsorbent particles, which can penetrate the adsorbate at elevated temperatures. Accordingly, effective interaction between adsorbate molecules and adsorbents increases with an increase in temperature. The percentage removal increases with the temperature, indicating that the adsorption mechanism is an endothermic phenomenon. Thus, removal of COD, BOD, and color using naturally prepared adsorbent is considered as chemical adsorption in which the adsorbate (components contributing to COD, BOD, and color) undergoes chemical interaction with naturally prepared adsorbent.

The effect of pH 3-11 on the adsorption by a–NLP, a–GLP, and TSP (dose: 5.0 g/L) for the removal of COD, BOD, and color, respectively, from dyeing wastewater at constant contact duration of 180 min, temperature of 300 K, and agitator speed of 400 rpm is presented in Table 4.5.

The percentage removals increase as the pH of the system increases. At pH 11, the highest percentage removals of COD, BOD, and color are found to be 64.8, 57.4, and 56.5; 60.0, 48.7, and 44.7; and 57.1, 42.8, and 40.5 by a–NLP, a–GLP, and TSP, respectively. pH is an important parameter in adsorption studies, but due to the presence of various elements in dyeing mill wastewater such as starches, dextrin, gums, glucose, waxes, pectin, alcohol, fatty acids, acetic acid, soap, detergents, sodium hydroxide, carbonates, sulfides, sulfites, chlorides, dyes, pigments, carboxymethyl cellulose, gelatin, peroxides, silicones, fluorocarbons, and resins, the moderate removal of COD, BOD, and color is found with a change in pH.

Table 4.6 presents the effect of different agitator speeds (200, 400, 500, 600, and 700 rpm) of the system on the removal of COD, BOD, and color of the wastewater at constant temperature (300 K) and a–NLP, a–GLP, and TSP (dose: 5 g/L). It can be seen that percentage removals increase as agitator speed increases up to 600 rpm; a straight line after 600 rpm indicates that equilibrium is attained at 600 rpm. At an agitator speed of 600 rpm, the highest percentage removal of COD, BOD, and color is found to be 52.5, 48.6, and 44.2, respectively, by a–NLP; 48.2, 40.1, and 36.8, respectively, by a–GLP; and 43.9, 38.0, and 31.1, respectively, by TSP.

Stirring is an important parameter in adsorption phenomenon, as, in the batch adsorption systems, agitation speed plays a significant role, affecting the external boundary film and the distribution of the solute in the bulk solution. The increase in ion exchange capacity can be explained by the fact that increasing stirring speed reduces the film boundary layer surrounding ions, thus increasing the external film transfer coefficient and the adsorption capacity. Stirring speed affects solution distribution on a solid–liquid system, so this parameter is very important for the adsorption phenomenon. With a further increase in agitation speed, some amount of dye adsorbed on the surface of the shell is desorbed due to the centrifugal force. Moreover, at high agitator speed, the adsorbed dye experiences a strong centrifugal force, which causes the loosely bound adsorbate to desorb from the naturally prepared materials. This is because agitation facilitates proper contact between the metal ions in solution and the biomass binding sites and thereby

Table 4.5 Influence of pH for the Removal of COD, BOD, and Color Using Naturally Prepared Materials

pH	a-NLP			a-GLP			TSP		
	COD	BOD	Color	COD	BOD	Color	COD	BOD	Color
3	14.8	9.7	2.9	7.5	3.2	1.4	8.0	5.7	3.7
5	23.6	17.6	11.3	14.8	10.2	7.1	24.2	21.7	14.2
7	39.0	32.1	22.9	26.5	20.2	11.4	34.8	31.7	28.2
9	56.5	50.1	44.2	42.1	34.7	28.2	50.4	40.0	34.0
11	64.8	60.0	57.1	57.4	48.7	42.8	56.5	44.7	40.5

Percentage removal

Table 4.6 Effect of Agitator Speed on the Removal of COD, BOD, and Color Using Naturally Prepared Materials

Agitator speed (rpm)	a-NLP			a-GLP			TSP		
	COD	BOD	Color	COD	BOD	Color	COD	BOD	Color
200	18.3	17.6	14.3	14.0	10.2	9.1	10.2	9.1	4.8
400	42.4	40.0	34.3	27.7	22.1	19.9	20.9	18.8	14.5
500	48.8	45.6	40.0	39.9	33.5	29.9	29.9	26.7	24.2
600	52.5	50.0	44.2	48.2	40.1	36.8	43.9	38.0	31.1
700	52.5	48.6	44.2	48.2	40.1	36.8	43.9	38.0	31.1

Percentage removal

promotes effective transfer of sorbate to the sorbent sites. Also, the pores on the surface of the adsorbents were unrestricted at high stirring speed, making all the pores available for adsorption until equilibrium.

4.4 CONCLUSION

(1) a–NLP, a–GLP, and TSP are utilized for the removal of COD, BOD, and color present in dyeing mill wastewater. Study indicates that a–NLP is the best among the three adsorbents investigated for dyeing wastewater treatment.

(2) In this adsorption study, different process factors like adsorbent dose, contact duration, temperature, pH, and agitation speed are exploited using naturally prepared adsorbents, namely, a–NLP, a–GLP, and TSP, for the removal of COD, BOD, and color.

(3) Higher efficiency (adsorption per gram of adsorbent) of adsorption is obtained at lower doses of a–NLP, a–GLP, and TSP.

(4) Maximum percentage removal is obtained using the parameter adsorbent dose by keeping other parameters such as contact duration (90 min) and temperature (300 K) constant. At a dosage of 25 g/L, when a–NLP was used, the highest percentage removals of COD, BOD, and color were 100%, 98.9%, and 100%, respectively; when a–GLP was used, 84.3%, 79.9%, and 78.5%, respectively; and when TSP was used, 81.0%, 74.9%, and 71.4%, respectively.

(5) Freundlich and Langmuir adsorption isotherm models are suitable for the removal of COD, BOD, and color from the dyeing wastewater by adsorption process onto adsorbents, namely, a–NLP, a–GLP, and TSP.

(6) The highest adsorption capacities (θ_0) related to Langmuir isotherm are found to be 1.34, 0.82, and 0.31 mg/g for COD, BOD, and color, respectively, using a–NLP. Using a–GLP, the adsorption capacities are found to be 1.146, 0.782, and 0.288 mg/g, and using TSP, the adsorption capacities are found to be 1.444, 0.476, and 0.281 mg/g for COD, BOD, and color, respectively.

CHAPTER 5

Fixed-Bed Column Studies of Dyeing Mill Wastewater Treatment Using Naturally Prepared Adsorbents

Contents

Abstract

Continuous fixed-bed column study is carried out using naturally prepared adsorbents for the removal of COD, BOD, and color from textile wastewater in this chapter. The effects of parameters such as flow rate and bed height are exploited, in which adsorption efficiency increases with increasing bed height and decreases with increasing flow rate. The Thomas, Yoon-Nelson, and BDST models are applied to predict the breakthrough curves and to determine the characteristic parameters of the column useful for process design using nonlinear regression. The adsorption capacity and rate constant associated with each model for column adsorption are calculated and mentioned. The maximum adsorption capacity related to the Adams-Bohart model is found to be 725.7, 729.8, and 380.4 mg/g for COD, BOD, and color, respectively, at a flow rate of 5 mL/min and bed height of 30 cm.

Keywords: Column, Adsorption, Column model, Flow rate, Bed height.

LIST OF FIGURES

Figure 5.1 Mechanism of breakthrough curve for sorption process in fixed bed.

Figure 5.2 Schematic diagram of fixed-bed column for COD, BOD, and color removal onto naturally prepared adsorbents.

LIST OF TABLES

5.1 INTRODUCTION

Batch process is important for estimating kinetic and thermodynamic parameters for an adsorption reaction and the interaction of the contaminations of interest with the adsorbent. It gives fundamental information for a particular sorbent-sorbate pair in terms of sorption capacities and kinetics. Batch adsorption experiments are used easily in the laboratory for the treatment of a small volume of effluents, which are limited using simple agitated closed vessel tests but are less convenient to use on an industrial scale, where large volumes of wastewater are continuously generated. Industrial utilization of sorbents screened using batch adsorption studies usually involves fixed-bed sorption. In a fixed bed, the sorbate is continuously in contact with a given quantity of fresh sorbent, thus providing the required concentration gradient between sorbent and sorbate for sorption.

Fixed-bed operations are widely used in pollution control processes such as removing toxic organic compounds and separating ions by an ion exchange bed by adsorption. In a batch operation, the adsorbent and adsorbate are in contact for a period of time until equilibrium is reached. In a column operation, adsorbate continuously enters and leaves the column; therefore, equilibrium is never achieved at any stage. As the solution flows down the column, the feed zone, which is the upper part of the packed adsorbent, will be saturated and the low concentration of adsorbate will encounter fresh adsorbent material at the bottom of the packing. The overall performance of the column is judged by its service time, which can be defined as the time the adsorbed adsorbate breaks through the column bed and is detected in the effluent. At that time, the column is considered to be saturated and column operation can be stopped. Continuous operation is the most suitable mode from both an economical and process control point of view. The continuous sorption process is usually characterized by the so-called breakthrough curves, that is, a representation of the pollutant effluent concentration versus time profile in a fixed-bed column. The major difference between batch and fixed-bed operations is in equilibrium establishment.

Column adsorption is more practical and more efficient to remove pollutants from real wastewater using fixed-bed columns. Fixed-bed operations have been widely used in various chemical industries for removing pollutants because of their simple operation and good adsorption capacity. A fixed-bed continuous flow column is an effective process for cyclic adsorption/desorption, as it makes the best use of the concentration difference driving force for adsorption, allows for a more efficient utilization of the adsorbent capacity, and results in better quality of the effluent. The performance of the fixed-bed columns in terms of a priori design and their optimization are described using the breakthrough curve concept. The breakthrough curve illustrates the behavior of a fixed-bed column from the point of view of the pollutant quantity that can be retained and is usually expressed in terms of a normalized concentration defined as the ratio of the effluent metal concentration to inlet concentration, as a function of flow time or volume of the effluent for the fixed-bed depth. The breakthrough curve is a representation of the pollutant effluent concentration versus time profile in a fixed-bed column. In addition, breakthrough curve prediction through mathematical models is a useful tool for scale-up and design purposes.

In a column, a solution is passed through a bed of sorbent beads where its composition is changed by sorption. The composition of the effluent and its change with time depend on the properties of the sorbent, the composition of the feed, and the operating conditions (flow rate, bed height, etc.). As the wastewater is fed into the column, most of the mass transfer takes place near the inlet of the bed, where the fluid first comes in contact with the sorbent. If the solid contains no sorbate at the start, the concentration in the fluid drops exponentially to zero before the end of the bed is reached. This concentration profile as well as the breakthrough curve are shown in Figure 5.1.

As the process starts, the solid near the inlet is nearly saturated, and most of the mass transfer takes place farther from the inlet. The concentration gradient is S-shaped. The region where most of the change in concentration occurs is called the mass transfer zone. This is the real behavior of the mass transfer process in fixed beds. When the axial or radial mass transfer resistances are neglected, sorption occurs homogeneously, and this is the ideal case. In fact, mass transfer resistances can be minimized but not effectively eliminated. Comments about such phenomenon will be better detailed. The limits of the breakthrough curve are often taken as C/C_o values of 0.05–0.95, unless any other recommendation is fixed. This is the case with wastewater treatment of highly toxic sorbates. When the concentration reaches the limiting permissible value or zero value, it is considered the break point. The flow is stopped, the column is regenerated, and the inlet concentration is redirected to a fresh sorbent bed.

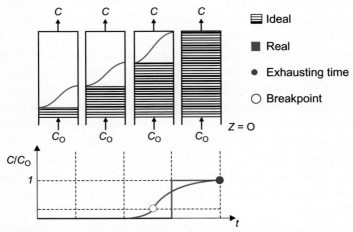

Figure 5.1 Mechanism of breakthrough curve for sorption process in fixed bed.

Adsorption dynamics acquaintance and modeling are essential because they provide valuable information on some practical aspects such as sorbent capacity and prediction of the time necessary for the effective operation of a fixed-bed column. At the same time, they assist in making more detailed conclusions about the mechanism of the process. Some of the mathematical correlations for adsorption in fixed-bed columns are based on the assumption of one mass transfer resistance, while others consider the influence of more mechanisms as well as the effect of axial dispersion. Many researchers apply various mathematical models and their modifications to account for the influence of some system parameters (flow rate, bed height, etc.) on the breakthrough curve progress. Mechanistic models for fixed-bed adsorption columns should include different phenomena such as axial dispersion, film diffusion resistance, intraparticle diffusion resistance (both pore diffusion and surface diffusion), and sorption equilibrium with the sorbent. The inclusion of all these processes involves rigorous but mathematically complex models associated with nonlinear partial differential equations. This implies using numerical solutions and, thus, a time-consuming process framework. In addition, the use of comprehensive mechanistic models involves the adjustment of several parameters; that is, the availability of several sets of reliable and well-designed experimental data is required. For this reason, in recent decades, a wide variety of semi-empirical models have been stated in order to predict the behavior of the fixed-bed adsorption process through the simulation of the breakthrough curve response.

5.2 EXPERIMENTS

Continuous adsorption of dyeing wastewater is carried out in a glass column with an internal diameter of 2 cm. The column is provided with five sampling points at 5 cm intervals. At the bottom of the column, glass beads with a 2 cm high layer are used to ensure uniform inlet flow to the column. Each adsorbent is filled with these glass beads to a height of 25 cm. The column is packed separately each time with 18.2, 19.5, and 20.1 g using a-NLP, a-GLP, and TSP, respectively. The wastewater is introduced into the column in bottom to top mode using a peristaltic pump at the desired flow rate. A schematic diagram of the fixed-bed column used in the adsorption study is shown in Figure 5.2.

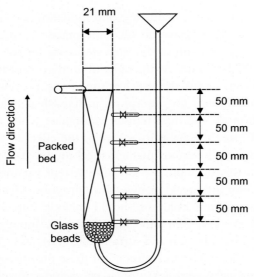

21 mm

Flow direction

Packed bed

Glass beads

50 mm
50 mm
50 mm
50 mm
50 mm

Figure 5.2 Schematic diagram of fixed-bed column for COD, BOD, and color removal onto naturally prepared adsorbents.

The wastewater is passed through the column with a controlled flow rate (5–15 mL/min). The neutral pH of the feed solution is adjusted by adding 0.1 N HCl or 0.1 N NaOH during the experiment. All chemicals used are of analytical reagent grade and purchased from Qualigens Fine Chemicals, India. Samples are collected periodically from all the sample ports and analyzed for residual COD, BOD, and color using standard methods, as mentioned in Chapter 2.

5.3 RESULTS AND DISCUSSION

The column adsorption experiment is carried out at different flow rates of 5, 10, 15, and 20 mL/min and a bed height of 30 cm using a-NLP, a–GLP, and TSP. The resultant breakthrough curves are shown in Figure 5.3a–c for the removal of COD, BOD, and color, respectively, at different flow rates (5, 10, 15, and 20 mL/min), in which breakthrough occurred faster with the higher flow rate of 20 mL/min. The breakthrough curve of the lower flow rate of 5 mL/min tended to be more gradual, meaning it is difficult to exhaust the column completely in all the three cases. The breakthrough time and the exhausting time at a flow rate of 20 and 5 mL/min are increasing from 165 to 390 min and 615 to 825 min, respectively, for COD; from

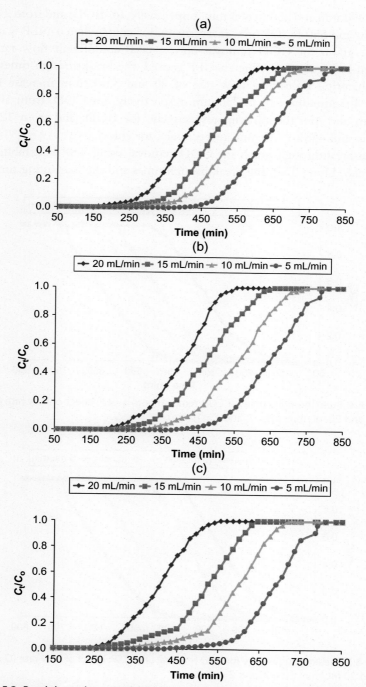

Figure 5.3 Breakthrough curves for (a) COD, (b) BOD, and (c) color removal onto a-NLP: effect of flow rate ($Q = 5$, 10, 15, and 20 mL/min, pH $= 7$, and $Z = 10$ cm).

195 to 390 min and 550 to 805 min, respectively, for BOD; and from 240 to 435 min and 550 to 825 min, respectively, for color when a–NLP is used.

The graph for COD removal onto a–GLP at different flow rates is revealed in Figure 5.4. When a–GLP is used, the breakthrough times and the exhausting times at a flow rate of 20 and 5 mL/min increase from 105 to 405 min and 510 to 735 min, respectively, for COD; from 105 to 330 min and 450 to 660 min, respectively, for BOD; and from 120 to 375 min and 435 to 735 min, respectively, for color.

The breakthrough graph for COD removal using TSP is exhibited in Figure 5.5. Using TSP, the breakthrough times and the exhausting times at

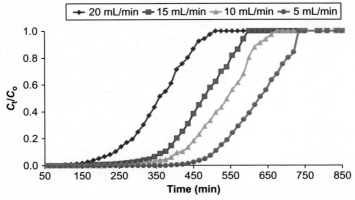

Figure 5.4 Breakthrough curve for COD removal onto a-GLP: effect of flow rate ($Q=5$, 10, 15, and 20 mL/min, pH$=7$, and $Z=10$ cm).

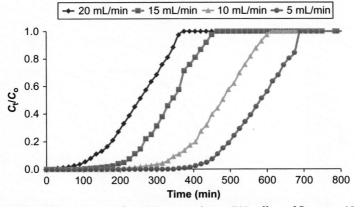

Figure 5.5 Breakthrough curve for COD removal onto TSP: effect of flow rate ($Q=5$, 10, 15, and 20 mL/min, pH$=7$, and $Z=10$ cm).

a flow rate of 20 and 5 mL/min increase from 30 to 360 min and 375 to 690 min, respectively, for COD; from 30 to 240 min and 375 to 535 min, respectively, for BOD; and from 15 to 240 min and 330 to 615 min, respectively, for color.

The column is found to perform better at a lower flow rate, which resulted in longer breakthrough and exhaustion times; that is, an early breakthrough is observed at the highest flow rate (20 mL/min), whereas the lowest flow rate (5 mL/min) exhibited a longer retention time. With a lower flow rate, the contact time between the sorbent and sorbate will be longer; thus, more sorbate can be retained within this interaction period, facilitating insufficient contact time to occupy the space within the particles. This behavior can be explained in terms of residence/contact time of the wastewater in the column. Less residence time is experienced by the wastewater as the flow rate increases. This will result in insufficient residence time for the diffusion processes of dye molecules into the pores of the naturally prepared adsorbents and limits the number of active sites available for adsorption, thus reducing the components contributing pollutants such as COD, BOD, and color to be treated. Also, at a lower flow rate, the residence time of the adsorbate is higher, and hence the adsorbent gets more time to bond efficiently. The liquid volumetric flow rate affects the rate of change in bed capacity in two ways: A higher liquid volumetric flow rate decreases the external film mass resistance at the surface of the adsorbents because of the additional velocity shear, thus reducing the film thickness. At the same moment, the residence time of the effluent inside the bed decreases with a higher liquid volumetric flow rate. Then, the adsorbate molecules have less time to penetrate and diffuse into the center of the adsorbent. Although the preceding feature is interesting from the industrial point of view, the increase in the flow rate must be balanced against the rise in the operating cost due to the larger pressure drop and pumping costs.

Thomas parameters [obtained from Thomas plot, i.e., $\ln((C_t/C_o) - 1)$ vs. t] like rate constant, K_{TH} (mL/(mg min)), and adsorption capacity, q_o (mg/g); Yoon–Nelson parameters [obtained from Yoon–Nelson plot, i.e., $\ln(C_t/(C_o - C_t))$ vs. t] like rate constant, K_{YN} (min^{-1}), 50% breakthrough time, $t_{1/2}$ (min), and adsorption capacity, Q_{oYN} (mg/g); and Adams–Bohart parameters [obtained from Adams–Bohart plot, i.e., $\ln(C_t/C_o)$ vs. t] like rate constant, K_{AB} (L/(mg min)), and adsorption capacity, N_o (mg/g), for COD, BOD, and color removal by column adsorption onto a–NLP, a–GLP, and TSP at different flow rates (5, 10, 15, and 20 mL/min) are shown in Table 5.1, respectively. Also, their correlation coefficient values (R^2) are calculated and mentioned.

Table 5.1 Thomas, Yoon-Nelson, and Adams-Bohart Parameters for COD, BOD, and Color Removal by Column Adsorption onto (a) a-NLP at Different Flow Rates, (b) a-GLP at Different Flow Rates, and (c) TSP at Different Flow Rates

| | | Column adsorption model | | | | | | | | | |
| | | Thomas | | | Yoon-Nelson | | | | Adams-Bohart | | |
Particular	Flow rate (mL/min)	K_{TH}, mL/(mg min)	q_o, mg/g	R^2	K_{YN}, 1/min	$t_{1/2}$, min	$Q_{o(YN)}$, mg/g	R^2	$K_{AB} \times 10^4$, L/(mg min)	N_o, mg/L	R^2
a-NLP at different flow rates											
COD	5	−0.1080	27.14	0.9846	0.0206	630.9	26.85	0.9846	25.49	725.69	0.8536
	10	−0.0951	45.71	0.9865	0.0190	537.1	45.91	0.9866	23.20	481.88	0.8971
	15	−0.0919	62.39	0.9859	0.0184	488.7	62.40	0.9860	22.33	198.74	0.9075
	20	−0.0910	70.74	0.9864	0.0182	415.6	70.74	0.9865	21.65	87.36	0.8952
BOD	5	−0.937	27.38	0.9887	0.0195	643.4	27.38	0.9886	25.52	729.79	0.9280
	10	−0.0946	46.57	0.9794	0.0189	547.2	46.57	0.9795	22.67	493.77	0.8891
	15	−0.0977	59.79	0.9772	0.0195	468.3	59.79	0.9973	23.23	190.50	0.8732
	20	−0.1117	67.33	0.9789	0.0223	395.6	67.33	0.9790	25.91	81.02	0.8963
Color	5	0.0945	21.54	0.9783	0.0118	680.4	28.95	0.9782	30.04	749.11	0.9863
	10	−0.0950	49.59	0.9623	0.0182	609.1	51.84	0.9622	24.45	504.05	0.9818
	15	−0.0923	65.13	0.9653	0.0185	510.2	65.13	0.9654	23.81	189.54	0.9754
	20	−0.1339	68.46	0.9821	0.0268	402.2	68.46	0.9822	29.40	80.85	0.8485

a-GLP at different flow rates

COD	5	−0.1066	26.67	0.9798	0.0213	626.6	26.67	0.9799	25.77	723.23	0.9124
	10	−0.0995	44.60	0.9896	0.0199	524.1	44.60	0.9896	22.76	479.83	0.9587
	15	−0.0916	60.89	0.9899	0.0183	477.0	60.89	0.9898	21.98	197.441	0.9811
	20	−0.0993	61.65	0.9908	0.0199	350.4	59.65	0.9908	15.88	84.64	0.9424
BOD	5	−0.0122	23.89	0.9751	0.0124	561.45	23.89	0.9752	21.72	697.51	0.9888
	10	−0.0984	41.07	0.9804	0.0197	482.58	41.07	0.9805	17.34	476.19	0.9914
	15	−0.0905	51.928	0.9453	0.0181	406.78	51.93	0.9454	16.51	187.99	0.9692
	20	−0.1112	55.59	0.9597	0.0222	326.59	55.59	0.9598	15.87	78.19	0.9707
Color	5	−0.1350	51.71	0.9598	0.0189	599.49	25.51	0.9599	34.11	676.20	0.8640
	10	−0.1091	56.51	0.9649	0.0201	533.25	45.38	0.9649	28.63	478.6	0.9335
	15	−0.1083	41.55	0.9809	0.0204	442.67	56.51	0.9809	25.33	170.17	0.9120
	20	−0.0947	25.51	0.9803	0.0270	303.81	51.71	0.9802	22.46	62.02	0.8922

TSP at different flow rates

COD	5	−0.1182	24.77	0.9821	0.0237	581.98	24.77	0.9820	32.08	638.49	0.9129
	10	−0.1056	39.89	0.9884	0.0199	468.76	39.89	0.9884	30.69	410.59	0.9505
	15	−0.1036	41.76	0.9911	0.0136	332.81	42.49	0.9910	27.85	134.12	0.9444
	20	−0.0997	42.49	0.9791	0.0211	245.37	41.76	0.9790	27.58	53.22	0.9375
BOD	5	−0.1445	17.99	0.9762	0.0289	422.79	17.99	0.9761	38.35	482.09	0.9774
	10	−0.1087	33.13	0.9611	0.0257	389.27	33.13	0.9610	30.68	339.11	0.9950
	15	−0.1034	39.24	0.9431	0.0217	307.35	39.24	0.9431	28.44	128.19	0.9682
	20	−0.1003	43.21	0.9549	0.0207	253.86	43.21	0.9549	26.79	54.84	0.9662
Color	5	−0.1312	19.69	0.9588	0.0181	462.9	19.69	0.9587	22.94	540.49	0.8681
	10	−0.1094	31.39	0.9690	0.0192	392.51	33.41	0.9691	25.34	353.98	0.9227
	15	−0.1023	41.19	0.9806	0.0205	322.69	35.26	0.9805	28.74	130.09	0.9110
	20	−0.0955	35.26	0.9642	0.0262	207.17	41.19	0.9641	33.96	45.62	0.9095

Adsorbent, a-NLP; flow rate, 5, 10, 15, and 20 mL/min; bed height, 10 cm.
Adsorbent, a-GLP; flow rate, 5, 10, 15, and 20 mL/min; bed height, 10 cm.
Adsorbent, TSP; flow rate, 5, 10, 15, and 20 mL/min; bed height, 10 cm.

It is observed that as the flow rate increases (5-20 mL/min), the values of K_{TH} (−0.1080 to −0.0910), K_{YN} (0.0182 to 0.0206), and K_{AB} (21.65×10^{-4} to 25.49×10^{-4}) increase, but q_o (70.74 to 27.14), $t_{1/2}$ (630.9 to 415.6), Q_{oYN} (70.70 to 26.85), and N_o (725.69 to 87.36) decrease for COD using a–NLP. Also, as the flow rate increases (5-20 mL/min), the values of K_{TH} (−0.1117 to −0.0937), K_{YN} (0.0189 to 0.0223), and K_{AB} (22.67×10^{-4} to 25.91×10^{-4}) increase, but q_o (67.33 to 27.38), $t_{1/2}$ (643.4 to 395.6), Q_{oYN} (67.33 to 27.38), and N_o (729.79 to 81.02) decrease for BOD when a–NLP is used. Further, as the flow rate increases (5-20 mL/min), the values of K_{TH} (−0.1339 to −0.0923), K_{YN} (0.0118 to 0.0268), and K_{AB} (23.81×10^{-4} to 30.04×10^{-4}) increase, but q_o (68.46 to 21.54), $t_{1/2}$ (680.4 to 402.2), Q_{oYN} (68.88 to 28.95), and N_o (380.41 to 80.85) decrease for color using a–NLP. The maximum adsorption capacity related to the Adams–Bohart model is found to be 725.7, 729.8, and 380.4 mg/g for COD, BOD, and color, respectively, at a flow rate of 5 mL/min and bed height of 20 cm when a–NLP is used.

In the case of a–GLP and TSP, as the flow rate increases (5-20 mL/min), the values of K_{TH}, K_{YN}, and K_{AB} increase, but q_o, $t_{1/2}$, Q_{oYN}, and N_o decrease for COD, BOD, and color derived from their respective graphs. At a flow rate of 5 mL/min and bed height of 20 cm, the maximum adsorption capacity related to the Adams–Bohart model is found to be 723.2, 697.5, and 278.6 mg/g for COD, BOD, and color, respectively, when a–GLP is used, and 638.5, 482.1, and 340.5 mg/g for COD, BOD, and color, respectively, when TSP is used.

The correlation coefficient values (R^2) for the Thomas, Yoon–Nelson, and Adams–Bohart models at different flow rate for the removal of COD, BOD, and color are greater than 0.93, suggesting the applicability of all these investigated models.

The effect of bed height for the removal of COD, BOD, and color from dyeing mill wastewater using the a–NLP bed at heights of 5, 10, 15, and 20 cm and a flow rate of 15 mL/min is mentioned in Figure 5.6a–c, respectively, which indicated that breakthrough times and exhaustion times for the bed heights of 5-20 cm are increasing from 195 to 450 min and 505 to 850 min, respectively, for COD; from 195 to 480 min and 540 to 865 min, respectively, for BOD; and from 255 to 480 min and 540 to 735 min, respectively, for color.

The effect of bed height for the removal of COD, BOD, and color from dyeing mill wastewater using the a–GLP bed at heights of 5, 10, 15, and 20 cm and a flow rate of 15 mL/min is drawn, which indicated that

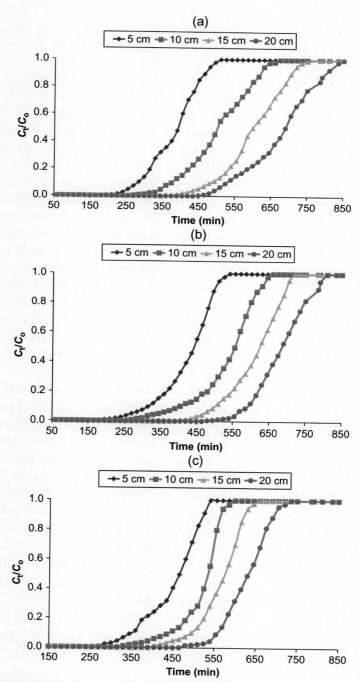

Figure 5.6 Breakthrough curves for (a) COD, (b) BOD, and (c) color removal onto a-NLP: effect of bed height ($Q = 15$ mL/min, pH $= 7$, and $Z = 5$, 10, 15, and 20 cm).

breakthrough times and exhaustion times for the bed heights of 5–20 cm are increasing from 90 to 330 min and 450 to 765 min, respectively, for COD; from 75 to 345 min and 420 to 600 min, respectively, for BOD; and from 105 to 375 min and 345 to 630 min, respectively, for color.

The effect of bed height for the removal of COD, BOD, and color from dyeing mill wastewater using the TSP bed at heights of 5, 10, 15, and 20 cm and a flow rate of 5 mL/min is also drawn, which indicated that breakthrough times and exhaustion times for the bed heights of 5–20 cm are increasing from 30 to 255 min and 420 to 660 min, respectively, for COD; from 90 to 240 min and 345 to 555 min, respectively, for BOD; and from 30 to 285 min and 285 to 540 min, respectively, for color.

An increase in the bed height (5–20 cm) increased both the breakthrough time and the saturation/exhaustion time. The breakthrough time at 50% of C_t/C_o, τ, also increased with an increase in bed height. This is because the zone of mass transfer had to travel farther from the entering point of the bed to the exit point. An increase in the bed height provides a greater number of fixation sites for binding by increasing the specific surface of the biomass and thus causing more sorption of substances containing pollutants, resulting in greater removals. The mass of the biosorbent also increased with an increase in bed height, which offered more surface area and adsorptive pores for the adsorption process and thereby additional space will be available for the adsorbate to be adsorbed on these unoccupied sites of adsorbent.

Initially, the feed solution is in contact with the fresh adsorbent at the bottom of the column. Components containing pollutants are adsorbed progressively on the sorbent as it flows upward. As more fluid is fed to the column, the bottom portion of the adsorbent becomes saturated with adsorbate; thus, the adsorption zone moves upward. Therefore, the concentration of the solute in the lower portions of the packed bed is usually higher than that in the top portions. The slope of the breakthrough curve decreases with increasing bed height, which results in a broadened mass transfer zone. A reduction in the bed height causes the predominance of axial dispersion rather than mass transfer; therefore, the concentration of solute did not have long enough for diffusion into the whole bed of sorbent, reducing solute diffusion. It is also evident that the breakthrough curves have the characteristic "S" shape, with a less pronounced shape in the case of the adsorbent mass of 0.1 g.

Thomas parameters [obtained from Thomas plot, i.e., $\ln((C_t/C_o) - 1)$ vs. t] like rate constant, K_{TH} (mL/(mg min)), and adsorption capacity, q_o (mg/g); Yoon-Nelson parameters [obtained from Yoon-Nelson plot, i.e., $\ln(C_t/(C_o - C_t))$ vs. t] like rate constant, K_{YN} (min^{-1}), 50% break-through time, $t_{1/2}$ (min), and adsorption capacity, Q_{oYN} (mg/g); and Adams–Bohart parameters [obtained from Adams–Bohart plot, i.e., $\ln(C_t/C_o)$ vs. t] like rate constant, K_{AB} (L/(mg min)), and adsorption capacity, N_o (mg/g), using a-NLP, a-GLP, and TSP for COD, BOD, and color removal by column adsorption at different bed heights (5, 10, 15, and 20 cm) and their correlation coefficient values (R^2) are calculated.

Parameters for column adsorption models (Thomas, Yoon–Nelson, and Bohart) for a-NLP for COD, BOD, and color removal are depicted in Table 5.2. It can be observed that as the bed height increases (5–20 cm), the values of K_{TH} (-0.0951 to -0.1325), K_{YN} (0.0270 to 0.0190), Q_{oYN} (57.41 to 32.24), and K_{AB} (34.11×10^{-4} to 23.21×10^{-4}) decrease, but q_o (32.83 to 57.01), $t_{1/2}$ (378.8 to 674.6), and N_o (287.89 to 670.66) increase for COD using a-NLP. It is further observed that as the bed height increases (5–20 cm), the values of K_{TH} (-0.0905 to -0.1246), K_{YN} (0.0249 to 0.0181), Q_{oYN} (58.01 to 35.45), and K_{AB} (34.49×10^{-4} to 23.43×10^{-4}) decrease, but q_o (35.05 to 58.23), $t_{1/2}$ (416.6 to 684.2), and N_o (279.3 to 639.8) increase for BOD using a-NLP. Further, as the bed height increases (5–20 cm), the values of K_{TH} (-0.1246 to -0.1633), K_{YN} (0.0327 to 0.0249), Q_{oYN} (53.24 to 38.44), and K_{AB} (38.86×10^{-4} to 35.00×10^{-4}) decrease, but q_o (38.34 to 53.13), $t_{1/2}$ (450.5 to 624.3), and N_o (78.9 to 310.9) increase for color using a-NLP. The maximum adsorption capacity related to the Adams–Bohart model was found to be 670.66, 639.8, and 310.9 mg/g for COD, BOD, and color, respectively, at a flow rate of 5 mL/min and bed height of 20 cm when a-NLP was used.

Also, as bed height increases (5–20 cm), the values of K_{TH}, K_{YN}, Q_{oYN}, and K_{AB} decrease, but q_o, $t_{1/2}$, and N_o increase for COD, BOD, and color using a-GLP and TSP. The maximum adsorption capacity related to the Adams–Bohart model is found to be 582.6, 554.9, and 315.9 mg/g for COD, BOD, and color, respectively, when a-GLP is used and 531.1, 469.6, and 301.4 mg/g for COD, BOD, and color, respectively, using TSP. These results are found at a flow rate of 5 mL/min and bed height of 20 cm.

The correlation coefficient values (R^2) for the Thomas, Yoon–Nelson, and Adams–Bohart models at different bed height for the removal of COD, BOD, and color are greater than 0.93, suggesting the applicability of all these investigated models.

The BDST parameters [rate constant, k (mL/(mg min)); adsorption capacity, N_o (mg/g); and correlation coefficient values, R^2] are mentioned

Table 5.2 Thomas, Yoon-Nelson, and Adams-Bohart Parameters for COD, BOD, and Color Removal by Column Adsorption onto a-NLP at Different Bed Heights

| | | Column adsorption model | | | | | | | | | |
| Particular | Bed height (cm) | Thomas | | | Yoon-Nelson | | | | Adams-Bohart | | |
		K_{TH}, mL/(mg min)	q_o, mg/g	R^2	K_{YN}, 1/min	$t_{1/2}$, min	$Q_{o(YN)}$, mg/g	R^2	$K_{AB} \times 10^4$, L/(mg min)	N_o, mg/L	R^2
COD	5	−0.0951	32.83	0.9902	0.0270	378.8	32.24	0.9595	34.11	287.89	0.8933
	10	−0.1004	42.59	0.9918	0.0229	500.5	42.60	0.9749	25.29	343.83	0.8870
	15	−0.1045	50.55	0.9895	0.0201	593.9	50.55	0.9800	24.56	445.84	0.9149
	20	−0.1325	57.41	0.9542	0.0190	674.6	57.41	0.9830	23.21	670.66	0.8433
BOD	5	−0.0905	35.45	0.9597	0.0181	416.6	35.45	0.9593	34.49	279.3	0.9707
	10	−0.1112	44.83	0.9453	0.0222	526.8	44.83	0.9454	30.69	336.9	0.9692
	15	−0.1122	51.92	0.9782	0.0224	610.1	51.92	0.9781	28.78	462.6	0.9416
	20	−0.1246	58.23	0.9939	0.0249	684.2	58.24	0.9940	23.43	739.8	0.9232
Color	5	−0.1246	38.34	0.9699	0.0249	450.5	38.34	0.9698	38.86	257.3	0.9164
	10	−0.1358	43.50	0.9131	0.0272	511.1	43.50	0.9130	36.69	310.9	0.9907
	15	−0.1517	47.54	0.9594	0.0303	558.6	47.56	0.9596	36.57	431.3	0.9805
	20	−0.1633	53.13	0.9849	0.0327	624.3	53.13	0.9848	35.00	749.6	0.9393

Adsorbent, a–NLP; flow rate, 10 mL/min; bed height, 5, 10, 15, and 20 cm.

in Table 5.2, in which the value of constant, k, decreases and N_o increases with increasing ratio of C_t/C_o. The maximum adsorption capacity related to BDST is found to be 7.11, 5.43, and 6.05 mg/g for COD, BOD, and color, respectively, at $C_t/C_o = 0.4$ using a–NLP, whereas the minimum adsorption capacity related to BDST is found to be 4.03, 3.98, and 4.56 mg/g for COD, BOD, and color, respectively, at $C_t/C_o = 0.2$ when TSP is used. Correlation coefficient values are greater than 0.87, suggesting the data are fitted to all four models (Table 5.3).

5.4 CONCLUSION

(1) In this study, the fixed-bed columns of a–NLP, a–GLP, and TSP are used for the removal of COD, BOD, and color, in which the breakthrough curves for flow rate (5–20 mL/min) and bed height (5–20 cm) are plotted. For different flow-rate, breakthrough occurred faster with higher flow rate of 20 mL/min. A breakthrough curve of the lower flow rate of 5 mL/min tended to be more gradual, meaning that the column is difficult to be completely exhausted in all three cases. Breakthrough time and exhaustion time decreased with an increase in flow rate.

(2) The effect of bed height for the removal of COD, BOD, and color from dyeing mill wastewater using an a–NLP bed at heights of 5, 10, 15, and 20 cm and a flow rate of 15 mL/min is studied, which indicates breakthrough times and exhaustion times increase with increasing bed heights of 5–20 cm.

(3) The experimental data were applied to the Thomas, Yoon–Nelson, BDST, and Adams–Bohart models. The observed linearity value (R^2) suggested that the data are fitted to all four models.

(4) It can be observed that as flow rate increases, the values of K_{TH} (rate constant of Thomas model), K_{YN} (rate constant of Yoon–Nelson model), and K_{AB} (rate constant of Adams–Bohart model) increase, but q_o (adsorption capacity of Thomas model), $t_{1/2}$ (50% breakthrough time), Q_{oYN} (adsorption capacity of Yoon–Nelson model), and N_o (adsorption capacity of Adams–Bohart model) decrease for COD, BOD, and color when a–NLP, a–GLP, and TSP are used. As the bed height increases, the values of K_{TH}, K_{YN}, Q_{oYN}, and K_{AB} decrease, but q_o, $t_{1/2}$, and N_o increase for COD, BOD, and color using a–NLP, a–GLP, and TSP.

Table 5.3 BDST Parameters for COD, BOD, and Color Removal by Column Adsorption onto a-NLP, a-GLP, and TSP

Parameter	Constant	a-NLP C_t/C_o			a-GLP C_t/C_o			TSP C_t/C_o		
		0.2	0.4	0.4	0.2	0.4	0.6	0.2	0.4	0.6
COD	N_o, mg/g	5.88	6.56	7.11	5.19	5.88	6.90	4.03	4.98	5.11
	$k \times 10^3$, mL/(mg min)	29.49	7.72	−6.84	44.72	11.49	−10.27	45.45	12.29	−9.13
	R^2	0.9892	0.9916	0.8470	0.9963	0.9889	0.9790	0.9794	0.9919	0.9991
BOD	N_o, mg/g	4.32	4.62	5.43	3.30	3.93	4.11	4.02	4.05	4.22
	$k \times 10^3$, mL/(mg min)	28.58	7.37	−7.44	30.14	8.45	−7.72	51.34	10.96	−10.04
	R^2	0.9884	0.9680	0.9840	0.9902	0.9764	0.9829	0.9791	0.9995	0.9961
Color	N_o, mg/g	5.13	5.88	6.05	5.10	5.46	5.67	4.56	4.62	5.01
	$k \times 10^3$, mL/(mg min)	44.01	11.42	9.84	55.45	13.52	−11.26	95.61	18.02	−15.02
	R^2	0.9703	0.9916	0.9940	0.9757	0.9841	0.9843	0.9782	0.9833	0.9667

(5) The maximum adsorption capacity related to the Adams–Bohart model is found to be 725.7, 729.8, and 380.4 mg/g for COD, BOD, and color, respectively, at a flow rate of 5 mL/min and bed height of 20 cm.

(6) The BDST parameters such as rate constant, k, and adsorption capacity, N_o, were mentioned, in which the value of constant, k, decreases and N_o increases with the increasing ratio of C_t/C_o.

(7) The removal of COD, BOD, and color using different adsorbents was found in decreasing order.

CHAPTER 6

Use of Naturally Prepared Coagulants for the Treatment of Wastewater from Dyeing Mills

Contents

Abstract

The wastewater generated by the textile industry is rated as the most polluting among all industrial sectors considering both volumes discharged and effluent composition. The present investigation is intended for the removal of COD, BOD, and color from textile wastewater using naturally prepared coagulants, namely, Surjana seed powder (SSP), maize seed powder (MSP), and chitosan. The effects of coagulant dose, flocculation time, and temperature are studied. The sludge volume index (SVI) and turbidity are examined for their various effects. SSP is more effective than chitosan and MSP for the removal of COD and color, and chitosan is more efficient than SSP and MSP in terms of SVI and turbidity. Maximum percentage reduction corresponding to 75.6 and 62.8 is obtained for the removal of COD and color, respectively, using SSP.

Keywords: Textile wastewater, Naturally prepared coagulants, COD, BOD, Color.

Characterization and Treatment of Textile Wastewater
http://dx.doi.org/10.1016/B978-0-12-802326-6.00006-X

LIST OF FIGURES

LIST OF TABLES

6.1 INTRODUCTION

In view of the growing awareness of pollution problems, dispersal of organic contamination in the environment is becoming a matter of concern. The ever-increasing use of chemical and related compounds in each and every field of industry and agriculture summons an urgent need for a method for their effective removal from water and wastewater. It is known that clay minerals and ore materials contain some active compounds that can be proved useful through various phenomena such as adsorption, coagulation, chemical precipitation, and flocculation in wastewater effluent treatment and thereby can contribute to the reduction of the wastewater problem. Clay minerals make an appreciable contribution to natural soils. They have an important function in natural water systems because of their large surface area per unit weight. In effect, clay minerals may operate as adsorbents for dissolved chemicals in water. Therefore, dissolved or suspended constituents in natural water or wastewater infiltrating soil horizons may react with clay minerals through adsorption, or clay minerals may be used as adsorbents in the treatment of wastewater. Researchers have been focusing their attention on studying physicochemical methods such as coagulant, an economic and viable method due to its highly selective nature, as alternate treatments for dyeing mill wastewater. Some naturally prepared coagulants for contamination removal have been investigated.[1,2]

The present study is intended to remove COD, BOD, and color from dyeing mill wastewater using naturally prepared coagulants, namely, Surjana seed powder (SSP), maize seed powder (MSP), and chitosan. The effects of various parameters like coagulant dose, flocculation time, and temperature are investigated. The sludge volume index (SVI) and turbidity are evaluated for various parameters.

6.2 MATERIAL AND METHODS

6.2.1 Surjana seed powder

Surjana seed is a seed of *Moringa oleifera*, a tropical plant belonging to the family *Moringaceae*. *Moringa oleifera* is one of the most widespread species that grow quickly at low altitudes; it is generally used as a vegetable, medicine, and source of vegetable oil. Surjana seeds are easily available in the Indian region. The mature seeds of the plant are washed with water to remove dust and are dried in an oven at 60 ± 2 °C. The dried seeds are crushed and powdered, sieved through 200 μm nylon sieves, and used as coagulant. The removal of dye, surfactant, and other contaminations using *Moringa oleifera* seeds has been studied by some scientists.[3–9] The chemical composition and properties of the active agent of *Moringa oleifera* Lam. seeds are demonstrated by Ndabigenge and Gassenscidit.[10,11]

6.2.2 Maize seed powder

Maize (*Zea mays*) is commonly known as corn in some countries. Similar to Surjana seeds, maize seeds are easily available in the Indian region. The mature seeds of the plant are washed with water to remove dust and are dried in an oven at 60 ± 2 °C. The dried seeds are crushed and powdered, sieved through 200 μm nylon sieves, and used as coagulant. Researchers had tried to utilize maize seeds as coagulant/coagulant aid.[12–14]

6.2.3 Chitosan

Chitin is a cellulose-like biopolymer widely distributed in nature, especially in marine invertebrates, insects, fungi, and yeasts. Deacetylated chitin, poly-β-$(1 \rightarrow 4)$-N-acetyl-D-glcosamine, is readily soluble in acidic solutions, which makes it more available for applications. Chitosan is a yellow powder and is a biodegradable, nontoxic, linear cationic polymer of high molecular weight with a variety of applications including water treatment, chromatography, additives for cosmetics, textile treatment for antimicrobial activity, novel fibers for textiles, photographic papers, biodegradable films, biomedical devices, improvement of quality and shelf life of food, and microcapsule implants for controlled release in drug delivery. Also, it is utilized for the recovery of suspended solids in food-processing wastes from poultry, eggs, cheese, and vegetable operations. It is procured from Sigma-Aldrich, India. Many scientists have successfully utilized chitosan as coagulant,[15-18] adsorbent[19-22] and with adsorbent.[23]

6.2.4 Experimental design

For the treatment of the wastewater samples, the coagulants at the required dosages are added to a small portion of the wastewater sample, stirred well and kept in contact for the requisite flocculation time and at the desired temperature under investigation, and then filtered. Important physicochemical characteristics, namely, COD, BOD, and color, are determined before and after treatment using standard methods. The pH of the system is maintained by 1.0 N HCl or 1.0 N NaOH during experiments. All other chemicals used are of analytical reagent grade.

To determine the effects of different amounts of naturally prepared coagulants, namely, SSP, MSP, and chitosan, the wastewater is treated with 5.0 g/L, 10.0 g/L, 15 g/L, 20.0 g/L, 25.0 g/L, 30.0 g/L, and 35.0 g/L of each coagulant for a constant flocculation time (120 min) at constant temperature (300 K). Effect of flocculation time is studied by treating the wastewater with 20.0 g/L of SSP, MSP, and chitosan at constant temperature (300 K) for 15-150 min. To determine the effect of temperature, the wastewater is treated with 20.0 g/L of SSP, MSP, and chitosan for a constant flocculation time (120 min) at various temperatures (298, 303, 308, 313, 318, 323, and 328 K).

6.2.5 Sludge volume index

SVI is a very important factor that determines the control or rate of desludging, i.e., how much sludge is to be returned to the aeration basin and how much to take it out from the system. It actually serves as a very important empirical measurement that can be used as a guide to maintain sufficient concentration of activated sludge in the aeration basin whereby too much or too little can be considered detrimental to the system's overall health. SVI is an indication of sludge settleability in the final clarifier. It is a useful test that indicates changes in the sludge settling characteristics and quality. By definition, SVI is the volume of settled sludge in milliliters occupied by 1 g of dry sludge solids after 30 min of settling in a 1000 mL graduated cylinder or a settleometer. SVI is a key factor when it comes to the clarifier design so that a clear wastewater discharge can be obtained without significant carry-over of sludge. It is computed by dividing the result of the settling test in mL/L by the mixed liquor suspended solid (MLSS) concentration in mg/L in the tank and multiplying it by 1000.[24,25] The equation is as follows:

$$SVI = \frac{(\text{Settled Sluged Volume in mL/L after 30 min}) \times 1000}{(\text{MLSS, mg/L})} \qquad (6.1)$$

6.2.6 Turbidity

Turbidity is a measure of the degree to which the water loses its transparency due to the presence of suspended particulates. The more total suspended solids in the water, the murkier it seems and the higher the turbidity. Turbidity can be caused by soil erosion, waste discharge, urban runoff, bottom feeders like carp that stir up sediments, household pets playing in the water, and algal growth. Turbid waters become warmer as suspended particles absorb heat from sunlight, causing oxygen levels to fall (warm water holds less oxygen than cool water). Photosynthesis decreases with lesser light, resulting in even lower oxygen levels. The main impact is merely aesthetic: nobody likes the look of dirty water. In addition, it is essential to eliminate the turbidity of water in order to effectively disinfect it for drinking purposes. This adds some extra cost to the treatment of surface water supplies. The suspended particles also help the attachment of heavy metals and many other toxic organic compounds and pesticides. Turbidity is measured in nephelometric turbidity units (NTU). The instrument used for measuring it is called a nephelometer or turbidimeter, and it measures the intensity of light scattered at 90 ° as a beam of light passes through a water sample.[26,27] The turbidity analysis of dyeing mill wastewater after treatment is conducted using a turbidimeter.

6.3 RESULTS AND DISCUSSION

6.3.1 Effect of coagulant dose

Coagulation dose is one of the most important factors to consider in determining the optimum condition for the performance of coagulants in coagulation and flocculation. Essentially, insufficient dosage or overdosing would result in poor performance in flocculation. Therefore, it is significant to determine the optimum dosage in order to minimize the dosing cost and sludge formation and also to obtain the optimum performance in treatment. The effects of coagulant doses (5-30 g/L) on the removal of COD, BOD, and color from dyeing mill wastewater at a temperature of 300 K and flocculation time of 60 min using chitosan, SSP, and MSP are shown in Table 6.1, which shows that there is continuous removal with increases in coagulant doses up to 30.0 g/L, which may be due to increases of substantially of coagulant. The same/lower removal value of COD, BOD, and color after 30.0 g/L indicates that an optimum dose of 25.0 g/L is found for all coagulants investigated. The highest removal of COD, BOD, and color is found to be 70.3%, 67.9%, and 62.8%, respectively, using SSP; 68.8%, 58.9%, and 47.0%, respectively, using MSP; and 64.7%, 55.7%, and 42.8%, respectively, using chitosan. This may be due to the resuspension of solids at this concentration. Furthermore, the high concentrations (>25.0 g/L) of the coagulant may confer positive charges on the particle surface (a positive zeta potential), thus redispersing the particles.

6.3.2 Effect of flocculation time

The time of macrofloc formation (flocculation time) is an operating parameters that is given great consideration in any water treatment plant that involves coagulation-flocculation operations. Figures 6.1–6.3 represent the effect of flocculation time on the removal of COD, BOD, and color using SSP, MSP, and chitosan, respectively, at a temperature of 300 K and coagulant dosage of 20 g/L. The consistent increment of removals is revealed with increasing flocculation time up to 120 min, and thereafter, the percentage removal is decreased or no percentage removal is found. The optimum flocculation time is found to be 120 min. The highest removal of COD, BOD, and color is found to be 75.6%, 66.1%, and 52.8%, respectively, using SSP; 74.7%, 62.9%, and 48.9%, respectively, using MSP; and 69.6%, 60.1%, and 42.8%, respectively, using chitosan.

6.3.3 Effect of temperature

The effect of temperature on COD, BOD, and color removal using chitosan, SSP, and MSP is investigated at 298, 303, 308, 313, 323, and 328 K, and the data are shown in Table 6.2. Continuous removals are found with increasing temperature up to 323 K. The optimum temperature is attained

Table 6.1 Influence of Different Doses of SSP, MSP, and Chitosan on Physicochemical Characteristics of Combined Wastewater

Coagulant dose (g/L)	SSP			MSP			Chitosan		
	COD	BOD	Color	COD	BOD	Color	COD	BOD	Color
5.0	26.62	15.68	9.99	20.73	12.19	8.00	13.24	7.69	5.14
10.0	37.59	30.15	21.70	31.79	21.74	16.85	25.42	17.68	13.71
15.0	48.53	43.22	37.12	50.54	32.14	24.27	40.09	28.62	19.99
20.0	61.94	58.98	49.77	54.62	43.25	33.32	50.38	35.63	28.56
25.0	74.11	67.66	58.51	62.53	49.90	44.49	56.43	48.60	38.49
30.0	70.34	67.66	62.82	68.82	58.88	47.03	64.68	55.56	42.83
35.0	69.42	67.66	58.54	67.73	58.88	45.69	63.53	53.56	42.83

Temperature, 300 K; flocculation time, 60 min.

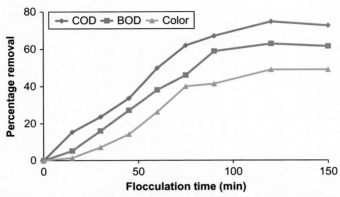

Figure 6.1 Influence of flocculation time on the removal of COD, BOD, and color using SSP.

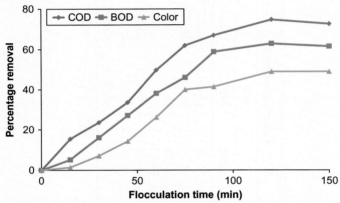

Figure 6.2 Influence of flocculation time on the removal of COD, BOD, and color using MSP.

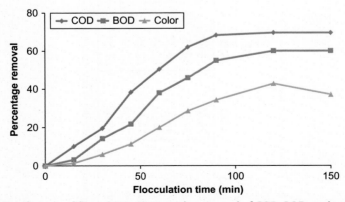

Figure 6.3 Influence of flocculation time on the removal of COD, BOD, and color using chitosan.

Table 6.2 Effect of Different Temperatures on the Removal of COD, BOD, and Color Using SSP, MSP, and Chitosan

Temperature (K)	SSP			MSP			Chitosan		
	COD	BOD	Color	COD	BOD	Color	COD	BOD	Color
298	19.3	14.2	8.6	24.2	10.2	5.4	18.1	5.1	3.1
303	33.8	28.8	17.6	31.6	18.8	14.3	26.8	9.2	5.7
308	48.8	40.1	28.6	43.7	25.7	20.0	36.1	18.7	11.4
313	61.9	51.4	40.0	56.5	38.0	26.6	48.6	29.0	22.6
318	67.1	63.6	55.2	63.2	49.9	42.9	57.4	52.7	42.9
323	74.9	71.1	67.1	72.6	68.1	58.5	69.6	64.6	47.3
328	74.9	71.1	67.1	72.6	68.1	58.5	69.6	64.6	47.3

Temperature, 300 K; flocculation time, 60 min.

Table 6.3 SVI and Turbidity Values of Investigated Parameters for Natural Coagulants

Effect of Factors	Chitosan SVI (mL/g)	Chitosan Turbidity (NTU)	SSP SVI (mL/g)	SSP Turbidity (NTU)	MSP SVI (mL/g)	MSP Turbidity (NTU)
Coagulant dose	312.2	325.5	425.5	400.1	541.1	512.1
Flocculation time	326.8	365.5	384.7	412.5	587.4	548.9
Temperature	354.5	387.8	397.8	467.3	599.9	587.8
Average	331.2	359.6	402.7	426.6	576.1	549.6

at 323 K for all coagulants. The highest removal of COD, BOD, and color is found to be 74.9%, 71.1%, and 67.1%, respectively, when SSP is used; 72.6%, 68.1%, and 58.5%, respectively, when MSP is used; and 69.6%, 64.6%, and 47.3%, respectively, when chitosan is used. At higher temperatures, higher percentage removals are achieved, perhaps due to better floc settlement when the temperature increases.

6.3.4 Analysis of SVI and turbidity

The stability of the microbial aggregates in activated sludge and the effluent quality are crucial in solid–liquid separation processes. Commonly, SVI is the most suitable factor to define the sludge settling ability. Table 6.3 mentions the average values of SVI and turbidity using various parameters like coagulant dose, flocculation time, and temperature for the removal of COD, BOD, and color from dyeing mill effluent. The decreasing order of SVI is found to be MSP > SSP > chitosan. It indicates that the lowest SVI is found more desirable by using chitosan. The average values of SVI for all parameters are found to be 331.2, 402.7, and 402.7 mL/g using chitosan, SSP, and MSP, respectively. So, higher values of SVI are found using MSP and then SSP and chitosan. Higher SVI values suggest poorer sludge compaction characteristics. The average values of turbidity for all factors (viz., coagulant dose, flocculation time, and temperature) including process optimization are found to be 359.6, 426.6, and 549.6 NTU using chitosan, SSP, and MSP, respectively.

6.4 CONCLUSION

(1)   The feasibility for treatment of dyeing mill wastewater using naturally prepared coagulants, namely, SSP, MSP, and chitosan, in order to remove COD, BOD, and color is analyzed.

(2)   A comparative study of these coagulants is conducted in which SSP is found to be more preferable than chitosan and MSP for the removal of COD, BOD, and color.

(3)   Various factors like coagulant dose, flocculation time, and temperature are investigated, in which coagulant dose is found to be more preferable than other parameters for the removal of COD and BOD. Also, flocculation time is more convenient than other parameters investigated for the removal of color.

(4)   The highest removal of COD (75.6%) and BOD (66.1%) is achieved using the coagulant dose of SSP (30.0 g/L) at a temperature of 300 K and flocculation time of 60 min. The highest color removal (62.8%) is attained using flocculation time (120 min) at a temperature of 300 K and SSP of 20 g/L.

(5)   The average maximum SVI and turbidity are found to be 402.7 mL/g and 549.6 NTU, respectively, using MSP and 331.2 mL/g and 359.6 NTU, respectively, using chitosan. Therefore, chitosan is more preferable than SSP and MSP.

REFERENCES

1. Oladoja NA, Aliu YD. Evaluation of plantain peelings ash extract as coagulant aid in the coagulation of colloidal particles in low pH aqua system. *Water Qual Res J Can* 2008;**43** (2/3):231–8.
2. Oladoja NA, Aliu YD. Snail shell as coagulant aid in the alum precipitation of malachite green from aqua system. *J Hazard Mater* 2009;**164**:1496–502.
3. Kwaambwa HM, Hellsing M, Rennie AR. Adsorption of a water treatment protein from *Moringa oleifera* seeds to a silicon oxide surface studied by neutron reflection. *Langmuir* 2010;**26**(6):3902–10.
4. Beltran-Heredia J, Martín JS. Azo dye removal by *Moringa oleifera* seed extract coagulation. *Color Technol* 2008;**124**:310–7.
5. Beltran-Heredia J, Sanchez-Martin J, Delgado-Regalado A. Removal of dyes by *Moringa oleifera* seed extract. Study through response surface methodology. *J Chem Technol Biotechnol* 2009;**84**:1653–9.
6. Bhatia S, Othman Z, Ahmad AL. Palm oil mill effluent pretreatment using *Moringa oleifera* seeds as an environmentally friendly coagulant: laboratory and pilot plant studies. *J Chem Technol Biotechnol* 2006;**81**:1852–8.
7. Vieira AMS, Vieira MF, Silva GF, Araujo AA, Fagundes-Klen MR, Veit MT, et al. Use of *Moringa oleifera* seed as a natural adsorbent for wastewater treatment. *Water Air Soil Pollut* 2010;**206**:273–81.
8. Beltran-Heredia J, Martın JS, Solera-Hernandez C. Anionic surfactants removal by natural coagulant/flocculant products. *Ind Eng Chem Res* 2009;**48**:5085–92.
9. Beltran-Heredia J, Martın JS, Solera-Hernandez C. Removal of carmine indigo dye with *Moringa oleifera* seed extract. *Ind Eng Chem Res* 2009;**48**:6512–20.
10. Ndabigengesere A, Narasiah KS, Talbot BG. Active agents and mechanism of coagulation of turbid waters using *Moringa oleifera*. *Water Res* 1995;**29**:703–10.

11. Gassenschmidt U, Jany KD, Tauscher B, Niebergall H. Isolation and characterization of a flocculating protein from *Moringa oleifera* Lam. *Biochim Biophys Acta* 1995;**1243**:477–81.
12. Raghuwanshi PK, Mandlol M, Shrama AJ, Malviya HS, Chudari S. Improving filtrate quality using agro based materials as coagulant aid. *Water Qual Res J Can* 2002;**37**:745–56.
13. Mandloi M, Chaudhari S, Folkard GK. Evaluation of natural coagulants for direct filtration. *Environ Technol* 2004;**25**(4):481–9.
14. Bhole AG. Relative evaluation of a few natural coagulants. *J Water Supply Res Technol AQUA* 1995;**44**(6):284–90.
15. Roussy J, Vooren MV, Guibal E. Influence of chitosan characteristics on coagulation and flocculation of organic suspensions. *J Polym Sci* 2005;**98**:2070–9.
16. Huang C, Chen Y. Coagulation of colloidal particles in water by chitosan. *J Chem Technol Biotechnol* 1996;**66**:221–32.
17. Chi FH, Cheng WP. Use of chitosan as coagulant to treat wastewater from milk processing plant. *J Polym Environ* 2006;**14**:411–7.
18. Szygla A, Guibal E, Palacın MA, Ruiz M, Sastre AM. Removal of an anionic dye (Acid Blue 92) by coagulation–flocculation using chitosan. *J Environ Manage* 2009;**90**:2979–86.
19. Mahmoodi NM, Salehi R, Arami M, Bahrami H. Dye removal from colored textile wastewater using chitosan in binary systems. *Desalination* 2011;**267**:64–72.
20. Minamisawa M, Minamisawa H, Yoshida S, Takai N. Adsorption behavior of heavy metals on biomaterials. *J Agric Food Chem* 2004;**52**:5606–11.
21. Gibbs G, Tobin JM, Guibal E. Influence of chitosan preprotonation on reactive black 5 sorption isotherms and kinetics. *Ind Eng Chem Res* 2004;**43**:1–11.
22. Sreelatha G, Ageetha V, Parmar J, Padmaja P. Equilibrium and kinetic studies on reactive dye adsorption using palm shell powder (an agrowaste) and chitosan. *Ind Eng Chem Res* 2011;**56**:35–42.
23. Huang C, Chen Y. Coagulation of colloidal particles in water by chitosan. *J Chem Technol Biotechnol* 1996;**66**:221–32.
24. Othman Z, Bhatia S, Ahmad A. Influence of the settleability parameters for palm oil mill effluent (POME) pretreatment by using *Moringa oleifera* seeds as an environmental friendly coagulant. In: International Conference on Environment 2008 (ICENV 2008), Penang, Malaysia, 15 to 17 December 2008; 2008.
25. Wong SS, Teng TT, Ahmada AL, Zuhairi A, Najafpour G. Treatment of pulp and paper mill wastewater by polyacrylamide (PAM) in polymer induced flocculation. *J Hazard Mater* 2006;**135**(1–3):378–88.
26. http://www.yokogawa.com/us/is/downloads/pdf/analytical/APPNOTES/TB_A_001.pdf.
27. http://www.lenntech.com/turbidity.htm.

INDEX

Note: Page numbers followed by *f* indicate figures and *t* indicate tables.

Printed in the United States
By Bookmasters